SOMETHING NEW?

AIs and us: the coming age when "nothing
new under the sun" is not true anymore

David Orban

Futuroid Press

To my wife, Diana, who anchors me to reality.

To my children Jacopo, Cosimo and Giordana who grew up hearing daily conversations about the topics covered here.

And to my mother Magda, who as a painter since I was a child, has oriented me in a scientific direction.

CONTENTS

PREFACE

The Technological Singularity is the hypothetical moment in which artificial intelligences will be able to modify themselves. They will be able to choose their own objectives and adopt the best methods to achieve them, thus changing the face of the planet and the course of history. What can we do? How should we adapt to this new reality?

It is tempting to align oneself with the wisdom of the nihil novum sub sole (there is nothing new under the sun) stance. But once in a while there are truly new things around us, and it is crucially important to recognize them and to be able to properly understand them. The forthcoming era of artificial intelligences is such a fundamental new component in our world.

This book covers the themes of technology, artificial intelligence and the evolution of the individual and society. I will gladly receive any kind of feedback or comment. You can contact me on one of the platforms listed below.

Email: david@davidorban.com
Web: davidorban.com
Twitter: twitter.com/davidorban
Facebook: facebook.com/searchingforthequestion
YouTube: youtube.com/davidorban

If you want to support my work, you can easily do so on Patreon at patreon.com/davidorban

TECHNOLOGY CREATED HUMANITY

A fiery start

There are many definitions of what a human being is, and what makes us different from other animals. That we are different is undeniable. Monkeys and many birds imitate and learn. Many human characteristics we might think are unique appear in other species. Yet they are combined in us in such quantities and in such a way as to have generated a qualitative change that makes us unique.

Among our various frames of reference, one is particularly useful: the technologies we have invented and applied not only help and support us in our lives, they define our nature.

One of the first examples of how technology has a fundamental influence on what we are is the ability to control fire. By using fire to cook the food we eat, we make fire an element in the digestive process. Our cousins the gorillas spend a dozen or more hours eating and digesting, whereas we are more efficient: we spend less time and energy eating and digesting, and also absorb the nutrients in the food more effectively. This has enabled us to shorten our digestive tract, and to expand our brain, notably the neocortex. Although the brain accounts for only about 2% of our body mass, it absorbs around 30% of our energy! The brain would grow even more were it not for the bottleneck, quite literally, represented by the diameter of the female pelvis. Even then, we are particularly immature at birth, compared with the newborn of other animal species. A few minutes after birth, a gazelle is able to stand up and run with its mother. A human baby is completely helpless and takes several years even to learn to feed itself. In this period, our brain continues to develop, especially the neocortex, the most recent part of the brain, organizing itself to accommodate as best as possible the huge quantities of information and knowledge needed to be an active and useful member of society.

Agriculture is another example of a key enabling technology. For tens of thousands of years, although human beings were equivalent to us today in every way, with the same characteristics and capabilities, the number of people living on the planet hardly changed. They were hunter-gatherers and the maximum size of these nomadic tribes was determined by the capacity of a particular area that could be covered by foot in a day to provide them with sustenance. The total population of the planet was only a few million. Indeed, recent genetic studies on the mitochondrion, the part of the cell whose genetic composition is only inherited through the mother, have shown that the number of people living at a certain point from whom all human beings are descended was about 5,000. This fine-mesh evolutionary sieve, and the chain of improbability it represents, is not just an isolated case: it represents a characteristic of natural selection and the evolution of complex systems. One reason for this is that a particular geographical area could feed only a relatively small number of individuals. A group of 20–30 people might stop in a valley for a few days or weeks, and then follow berries as they ripened, or the movements of animals. We have an idealized view of that period, given our lack of direct references. We imagine a sustainable lifestyle without commitments and stress, in touch with nature. In actual fact, this is only true if we accept an average life expectancy of 20–30 years, perhaps not even that, and the real unsustainability of a continual plundering of nature, which is only allowed to catch its breath when we move on. An illustration of the unsustainability of prehistoric man's way of life is hunting, which led to the extinction of the megafauna in all the continents. Whether it was the Siberian mammoth or the Australian emu, we killed them all without a second thought. So much for the image of the "noble savage" in contact with nature!

The advent of agriculture simultaneously in various regions of the world about ten thousand years ago, based on the cultivation of corn, rice and maize and the rearing of chickens, pigs, sheep, cows and other domestic animals, enabled man to increase the quantity of food available in a particular geographical area by more than an order of magnitude. The reliability of this production, even despite the variations in yields due to rainfall and parasites, was also preferable to the impossibility of predicting what a group of nomadic hunter-gatherers would find in the next valley. In turn, these developments led not only to the establishment of permanent settlements, but also to a corresponding increase in orders of magnitude of the population.

Paradoxically, statistical processing of the information acquired from skeletons of that time shows that in the agricultural age, life was harder than when people were hunter-gatherers: the average person was shorter and died younger! Even if people realized this, it was difficult to do anything about it: in the agricultural regions, it was no longer possible to return to the old lifestyle and social organization. This is a general principle of societies and organizations that adopt certain technologies: the new transformed reality depends on technology to function and can no longer abandon that technology and go back to the old ways. Our image of bygone days is often idealized; our perception of their positive aspects is amplified, as is our perception of the negative aspects of the present. Yet if we decided to abandon technology today, billions of people would die and no-one, luckily, is able to decide who those people should be.

The increased availability of food and its relative reliability in the agricultural age fueled a population increase and, with this increase, the development of specialized activities. Initially, these activities were directly associated with agricultural work, but, later, a growing percentage of the population became involved in non-agricultural occupations. Today, in higher-income societies, not more than 2–3% of the population works in agriculture and animal husbandry: we are able to feed 100% of the members of society with just 2% of the work force. Even in societies with medium-low income such as India, irrespective of the percentage of the population, it has been calculated that mechanical sources account for 90% of the energy involved in agricultural work.

The dinosaurs did not have telescopes

For tens of millions of years, dinosaurs were the cutting edge of evolution, with a great variety of species. We are still discovering how dinosaur groups were organized and, for example, how some of them cared for the young. We do not know what triggered the development of intelligence in man nor do we know whether dinosaurs would have made similar progress, given an appropriate environmental stimulus. We do know that time ran out on them: after the impact of an asteroid with the Earth, and the ensuing climatic changes, dinosaurs became extinct, together with a very large majority of contemporary species. There have been five major ones of these so-called mass extinctions on the Earth and we are investigating their respective causes.

If dinosaurs had had telescopes, could they have avoided extinction? They would certainly have been able to identify the asteroid before it hit the planet. And if they had had a sufficiently advanced space technology, they could have organized a rescue mission to try and modify the orbit of the asteroid and prevent it from colliding with the Earth. We have telescopes and are gradually developing our capabilities in space and the models enabling us to perform maneuvers of this type. For example, and despite what Hollywood movies would have us believe, exploding the asteroid is not particularly helpful; although the pieces would be a bit smaller, they would remain on the same orbit, and their total impact would have the same energy as the whole asteroid. The thinking today inclines toward the view that anchoring a sufficiently large probe to the asteroid would create a variation in the gravitational field that would change its orbit over time and avoid a collision with the Earth. The question involves mathematical and scientific knowledge, engineering expertise, project coordination and management, finance and resource allocation, and social and political consensus.

Figure 1: The only difference between us and the dinosaurs is that they didn't have telescopes.

In concrete terms, the difference between us and the dinosaurs, of course, is that they didn't have telescopes. Metaphorically, however, they did not possess the tools of reason and science to enable them to deal with the dangers of extinction, to see and perhaps deviate the asteroid, or, in our modern times, the dangers of pandemics, climate change, extreme conflict and so on.

When the NASA budget for radio telescopes that catalogue and monitor objects in space whose orbit could bring them into collision with our planet was cut, we turned ourselves into dinosaurs, voluntarily blind to the threats to our species. Microsoft co-founder Paul Allen had to step in, using his own resources to finance the re-activation and management of these specialized tools, and give us a chance to catalogue the asteroids and perhaps prevent a future impact from wiping us out.

Generally speaking, it seems fair to say there are no alternatives to the tools of reason and science to deal with our problems. Even when we put them to the best possible use, we have no guarantees. Insurmountable problems exist. But if we want to have a chance of resolving them, we simply have to identify the most appropriate scientific approach, the most suitable type of solution (for example, a gravitational solution rather than a Hollywood-style bomb). Burying our heads in the sand is not the answer.

A non-zero sum game

There is a common belief in the press, radio and television, but also among online media, websites and the social networks, that the only serious way of providing information is to present the facts in a balanced manner. Taken to an almost absurd extreme, this approach sees two sides to every fact or phenomenon, one positive, the other negative, and tries to give the same amount of column space or airtime to both. Whether this dogmatic method stems simply from ignorance or whether vested interests are not infrequently at play is hard to say. Among scientists, 99% are convinced that the climate is changing and that man is the cause. You often see television programs where an almost comic attempt is made to give equal time to the argument that this is not the case (that climate change does not exist or is not caused by man). By inviting a scientist on one side, and a climate change denier on the other, these programs create the artificial and false impression that both points of view are equally valid. Based on this type of approach, it is easy to make the flawed generalization that there are two sides to every technology, one positive and one negative.

Technology, however, is not a zero-sum game, but a positive-sum game. Rather than balance out, its positive and negative sides have a positive net effect. This valuation is statistical, not absolute: the various facets of each technology need to be carefully examined. After an open debate in which everyone takes part, we may decide not to adopt certain technologies, to avoid becoming dependent upon them. Generally speaking, however, the simple fact that the world population is more than seven billion rather than a few million demonstrates that the overall effect is beneficial to humanity. People who say we should abandon technology, because they fail to understand it and therefore fear it, should first of all answer the question "Who are the 99 people out of 100 who will have to die as a result?"

The precautionary principle that highlights the risks of technologies slows their uptake. Questions relating to consumer protection are often cited with regard to their use. The premise here is that on one side, the consumer is defenseless and unable to defend themselves, on the other that the consumer is under attack, ready to be exploited and tricked, and generally opposed to the action of those who propose solutions to their problems. The final step in this reasoning implies that the regulatory bodies are better informed, better prepared and better able to decide for the best on behalf of the consumer, in deciding what the consumer may or may not be able to do or even to know.

Access to the sacred text of your DNA

A recent example of this precautionary principle at work is the opposition of the US Food & Drug Administration, the federal agency that regulates food and healthcare products and services, to the DNA decoding service offered to the public by the 23andMe company.

Taking advantage of the exponential reduction in decoding costs, in 2009 this Californian company began offering the public an innovative service. When you register on their site, they send you a test-tube: you deposit a sample of your saliva in the tube (put bluntly, you spit into it) and mail it back to the company in the envelope you received in the original package, prominently marked with the threatening "biohazard" icons for biological materials. After a couple of weeks, they send you an email message informing you that the decoding procedure has been completed. You can then read the results, protected by a password, directly on the company's website.

In 1985, an extremely ambitious Human Genome Project was launched in the USA with a budget of three billion dollars. The goal was to decode the human genome, consisting of three billion base pairs, in fifteen years. The progress of the project is a classic example of the power of exponential change. Seven years in, only one percent of progress had been achieved. Even the experts in the field thought the project had failed and that it would take not another seven years, but dozens if not hundreds of years to complete, and involve astronomic costs many times higher than the funds originally provided. Very few people considered that the one percent result had been achieved through a doubling of the decoding capability. By maintaining this rhythm, not the speed, but the acceleration of the process, by the following year, 2% progress had been achieved, then 4%, 8%, 16%, 32%, 64%... and after exactly seven additional doublings, the 100% objective had been achieved on time – fifteen years – and on budget.

The doubling of the power of the project didn't stop there, however; the speed of decoding increased, bringing down its cost. Today, in 2015, it is possible to have a complete DNA profile for about $2000, or for $99 a partial profile, of the so-called Single Nucleotide Polymorphisms (or SNIPs) believed to be responsible for the individual characteristics that distinguish us from one another. There are five hundred thousand SNIPs in human DNA and the 23andMe company concentrates on their analysis and processing.

The results are astonishing. By analyzing my DNA, the people at 23andMe can tell me the color of my eyes and hair, as well as dozens of other characteristics of my phenotype; that is, the physical manifestation of the action of my DNA. They can make a statistical assessment, in terms of greater or lesser predisposition toward specific types of illness. They provide indications on dosages even for common drugs if I should have to use them, in addition to the recommendations of normal prescriptions, or, conversely, advise a lower dose given my natural reaction to the compounds contained in the medicine.

Clearly, being able to take decisions on the basis of this information can have important consequences. Knowing that specific changes in my lifestyle can reduce the probability of a particular condition affecting me can be life-changing. Telling my doctor that I have a particular sensitivity to a drug he is about to prescribe so he can adjust the dose accordingly could save my life.

The FDA has decided that the way 23andMe presents the information, making the probabilistic and statistical correlations between the genome, behavior and the development of particular conditions and diseases explicit, was not appropriate. Specifically, it has ruled that consumers should not have direct access to this information, which should be given only to physicians who would therefore be the only people to interpret the data and give advice to their patients based on their conclusions.

In 1517, Martin Luther nailed his Ninety-Five Theses on the door of All Saints' Church in Wittenberg. His symbolic act was the trigger for the development of the Protestant movement in the Christian church, leading to a schism with Catholicism that continues today. One of Luther's theses was that the Bible, the sacred text of Christianity, should be translated from Latin into German to enable people to read it themselves, without the intermediation of a priest. The preexisting interests, as well as a conservative and dogmatic justification, created an irresolvable conflict that led not only to the schism in the church, but also to hundreds of years of bloody conflict.

Today the FDA is taking the role played by the Vatican in Luther's day. It does not want people to have access to the sacred text of their DNA, translated from the language of biochemistry into the accessible language of information technology, and has ruled that interpretation of the text may only be through the priestly intermediation of medical practitioners, who uphold their position from a conservative and dogmatic basis.

The proactionary principle

Our actions shape the future. The consequences of our hopes and ambitions extend beyond the present.

The precautionary principle dictates that before a given solution is adopted, it is necessary to take into consideration all the harm that it can cause. Often invoked in areas of consumer protection, regulators feel empowered by it to make sure that new products and services that are brought to market are not only useful and have positive effects, but that negative effects can be excluded. Especially in the fields of health and pharmaceutical research, the cautionary principle has been the preeminent inspiration of product development.

Of course in an ideal world regulators would only seek to establish the best course of action for the public good, rather than implementing self-propagating bureaucracies. And in an ideal world dominant players in a given market would not use their power to stop newcomers and diminish the dangers of unknown competitive factors, distorting the rules, and unduly influencing the process. Yes, we are not living in an ideal world.

The proactionary principle, originally proposed by Max More, takes into account the opportunity cost of inaction, and the costs of regulation itself, in deriving a balance that is more future oriented. If we look at the future generations, and the benefit they derive from our actions, delaying the implementation of a new technology can have very large consequences.

The freedom to experiment, and the opportunity for individuals to gain knowledge outside of the officially sanctioned paths of research, objectivity, and transparency, and other components of the proactionary principle make it a useful tool to design novel action.

A family residing in the United States has been afflicted by a rare condition for which there wasn't a commercially available cure. The pharmaceutical company that actually started initial research on it evaluated that going through the regulatory obstacle course to bring it to market was not worth it. Of course for those directly impacted by the illness it is not a question of profit considerations. Thanks to modern communications, the availability of research, the family has been able to connect with others in their same situation, to purchase the rights to further develop the cure, and to successfully apply it to their own members and to those of other families similarly afflicted.

Based on the traditional approaches, this could never have happened, both from a technological and regulatory point of view. There are countless other examples of bolder experimentation and free inquiry that await the application of the proactionary principle to achieve their goals and flourish.

The European Union, based on environmental concerns, incorporated the precautionary principle in its fundamental treaties. Is it going to imply that the EU is more likely to refrain from adopting technologies than other socio-economic areas? Is this going to be the basis of a certain level of fossilization of European society? Or maybe it is the expression of this situation which already afflicts it?

Fit or unfit civilizations

Is freedom an emergent property of self-organizing matter? We strongly believe to be endowed of free will, and most of our social structures are based on this. There is no physical foundation to it. The determinism of physical laws, quantum uncertainties notwithstanding, allows no space for the concept to hide and show its effects. Just as we are constantly moved to anthropomorphizing objects, animals, and phenomena, we are compelled to interpret decisions as made freely instead of being the consequence of the state of matter and its interactions, inside and outside of us.

Individual behavior aggregates to that of larger groups, and finally of societies. We judge the outcomes of individual decisions and consequences in civil and criminal law. We can judge the capacity of societies to engender the well-being of its members, or, on the contrary, to be corrupt, unjust, and spreading confusion, violence and suffering.

The capacity of a group of societies to generate well-being doesn't only depend on the aggregate decisions of its members. It also depends on what knowledge is actually available to them, and through them to it. The ancient Roman civilization created wonderful art and philosophy, and we rightly admire its achievements. However, it was organized at a fundamental level on slave labor, which today is universally condemned. Could it be different? Is it possible to imagine a Roman civilization that didn't employ slavery? It is not, because the level of knowledge, and especially the energy availability that that knowledge generated, made it impossible to achieve its goals without resorting to the force of human muscle, or programmable humans, people you could tell what to do, and who'd do it without talking back.

During its period of expansion, while able to draw on successful waves of slaves, the Roman civilization appeared to be well adapted. It was only an appearance, because it could not last. Rome could not further expand, having basically conquered all the available land of what we call today Europe, North Africa and the Middle East, and enslaved all the individuals that could be enslaved in those populations. At that point it went into decline and became incapable of resisting the changes coming, or adapting to them. This is of course an extremely simplistic representation of a long and complex history. There are many other forces at play beyond that of slaves or the lack of slaves.

Today, with current knowledge, we can build societies and civilizations without slaves. Relying on chemical and soon solar energy, our decisions are driven by their more efficient use, and they outcompete the eventual alternatives. We are not morally superior to the Romans as individual human beings, we are taking advantage of the accumulated information and its applications. The outcome of the US Civil War between the South and the North was dictated by economic efficiency and a better organization of energy and industrial bases by the North.

Our current civilization is, as a consequence, the expression of our knowledge. The technologies we have available shape it, similarly to how Roman civilization was shaped by the knowledge and the technology available at the time. We can start asking ourselves what the limits of adaptability of our civilization are, and how it will change with the accumulation of information, its application in new knowledge and new technologies. If you asked a Roman to tell you if it were possible to build a civilization without slaves, the answer would have been "No!". What are the false axioms that we are holding? What are the questions that we can ask and assume that the answers would be universal, with everybody firmly believing that a given assumption is a necessary part of our societies in any place and any time? With the knowledge available to us in the future, we'll appear as primitive and naive with that secure and false answer as the Romans appear to us today.

When the shift happens, and how it manifests itself, depends on the tensions that build up between societies in a given era. What is possible in one place may not be immediately possible in another place. Differences build up as a consequence, since knowledge gets applied and experience accrues. In a world of global communication as today, the understanding of these differences brings to the possibility of applying knowledge faster, adopting best practices, what works well, and avoiding mistakes. When communication barriers or ideological ones put obstacles in the path of this flow of information, the divergence of societies increases. The tensions accumulate, and under an apparent immobility, the organizational structure of society is under increasing stress. At that point a small change in the boundary conditions can bring to very big core changes, rippling through the entire society.

This is what literally happened with the Berlin Wall, which was metaphorically and physically shielding the planned economies of the USSR and of Eastern Europe from those of the West. When the Wall fell, the effects of allowing market economies to rapidly penetrate brought first economical and then swift political change which could not be contained or controlled even by the very people who initiated and allowed them, like then Secretary General of the Russian Communist Party, Mikhail Gorbachev.

Since the information differences create these areas of limited knowledge, individuals within those areas often don't even realize that they are living in a maladapted society. They can be taken by surprise when the weaknesses and the brittleness of the civilization are made evident by the abrupt changes. Even political experts and historians are better at explaining rapid civilizational changes after the fact than forecasting them. This makes it hard to prepare for the changes, and to reduce the amount of suffering that the period of uncertainty creates.

Unsustainability is unsustainable

The current capitalistic economic paradigm predicated on

constant growth has been dominating for the past 200 years. Even before, resource exploration and allocation in feudal societies has allowed ignoring what today we call externalities of economic activities. This was possible while a given nation or civilization did not care about destroying a competing one. Or under the assumption that depleting the ecosystems and exterminating dominant species of a given continent there would always be another to discover and to start the process over with.

Today it is evident that this behavior is not possible anymore. Advanced nations should not wage war against each other. They should not destroy or subjugate other populations. They should not, endanger ecosystems and their species through their economic activities. Very simply there are no new continents available to pillage.

This implicitly means that we are not better people than before. We did not change our ways because we understood that there was a morally superior behavior to be adopted. Our mentality is fundamentally the same. The reason we are now considering alternatives is because the old way is not possible anymore.

The externalities of our economic activities are all those consequences that are not reflected in the profit and loss considerations. It is up to society to look at various sectors and decide if it can allow this. Alternatively, it can put in place regulations that surface the hidden costs, and let society as a whole explicitly absorb them, or force the enterprises active in the chain of production to deal with them instead.

Unsustainable economic practices have large externalities and in a closed and globally connected world cannnot be allowed. Overcoming the depletion of ecological support systems, the waste of resources that could be used more efficiently or recycled, a complex society must turn towards sustainable practices in order to generate dynamic, but robust solutions.

A civilization cannot be well-adapted without recognizing

this need, and without acting on it with the appropriate tools of incentives and regulations.

THE METHODS OF KNOWLEDGE

In order to survive, we must observe our environment, try to understand it, acquire resources we need, and plan actions in order to achieve our goals. Knowing what are the rules of the world, systematizing the knowledge and understanding how we can know better is useful to be better at surviving.

The fatal error of the alchemists

In a world that is dominated by competition and the universal perception is that resources are scarce, it is natural to choose a strategy of secrecy. Gathering knowledge in secret gives an advantage to those who can exploit it against others. A closed and well-guarded system of knowledge is a barrier that others have to overcome if they want to participate at the same level.

At the same time, a closed and secret system is also vulnerable to be isolated and suffering its own mistakes in isolation. Not being able to share its learnings, a community relying on secrecy is bound to be repeating mistakes because their lessons cannot be shared.

Medieval alchemists who were obsessed by the goal of transforming lead into gold were unlucky enough to believe that using mercury would help in this quest. Unfortunately mercury is poisonous. Any alchemist who learned this fact did it on their own, and suffered the consequences of this knowledge. The organization in secret societies blocked the possibility of learning from each other's mistakes, and the alchemists were bound to repeat them.

In today's world still there is a lot of activity that is conducted similarly in secret. It is assumed that sharing it would weaken the position of those who compete for resources.

Open science

The scientific revolution that Galilei started not only represented a clearer understanding of how theory and experiments needed to be related, it also paved the way for a profound shift in how knowledge and information was collected, checked, and spread. Without it being a necessary part of the scientific method itself, open collaboration allowed groups and individuals who took advantage of it to more rapidly decide if a certain set of results was reliable.

Open science is fundamentally superior to closed approaches of knowledge gathering. Collaboration among people who share the goals and passions of a given field is enhanced by a common language and tools. Publishing the results that one group achieves allows another group to run experiments against the new knowledge to confirm or disprove them. Interdisciplinary collaboration is enhanced by the ease with which practitioners outside of the group can approach it, building bridges of understanding that overcome specializations.

Today a fertile field of study made possible by the increasing digitalization of science and scientific publishing is that of meta-studies. Comparing and analyzing a large number of publications in a given field, it is possible to derive results not individually contained in any single one of them. Statistical tools reveal significant trends, and enhance the possibility of catching and eventually correcting methodological errors in the work previously published.

The world of science values facts, theories, verifiability, experiments, and the publication of results for sharing knowledge. As one of the main products of the scientist, if even not the sole one on which their advancement, capability to receive grants, and professorship are decided, scientific publications through the peer review system have been at the center of the development of science.

The value of the scientific article is measured through the number of other publications citing it, and the importance of the journal in which it appears. This has been converted directly in economic value by the publishers of the scientific journals, which have created sizable enterprises through charging universities and research institutions for the subscriptions to their periodicals in which the articles of the scientists appear. The price of these subscriptions has increased enough that universities in low and middle income countries or even some in high income ones are unable to afford the subscription price. There is also a more fundamental issue that publicly funded research and its results are subsidizing the revenues and the business models of private corporations who are collecting the articles and end up being paid twice over.

Open access publications have started to appear, and gain popularity and prestige, offering alternative models to scientific publication, where the reader of the articles in not charged either for single access or for a subscription to the journal. With the peer review system still in place for the quality control of the article, the author of the article is charged, and their institution, for the publication, a fee that is affordable and reasonable to be added as a simple line item to the budget of an experiment or the grant request.

With characteristic ruthlessness, science is also re-evaluating the effectiveness of the peer review system itself, trying to measure it on one hand, and on the other searching for possible alternative ways of assuring high levels of quality for scientific publications. The structure of scientific experiments is being analyzed in order to maximize the probability that the published results can be double checked through independent reproduction and verification of the data.

Making not only the scientific article itself, but lab notes and underlying raw data streams for further evaluation and aggregation has become the norm in several areas. With larger and larger data sets being generated in computer science, life science and many other fields, the analysis of the data has become a new burgeoning opportunity by itself.

Big data and data science offer the possibility to aggregate and sublimate value out of large sets of structured or unstructured information. Their application is becoming better understood in genetics, the Internet of Things, sociology and other fields. Cities and governments can on one hand make the data streams they generate available openly for others to leverage, in a permissionless, unencumbered manner where innovation and creativity can strive, and on the other hand they can take advantage of the results to increase transparency, accountability and efficiency of their operations.

The evolution of science

With the availability of increasing amounts of data, and the interconnectedness of the world and its experts, it is possible to be creative about the way that the pieces of the scientific puzzle are put together.

In the traditional process of undergraduate, graduate, doctorate and post-doctorate studies, increasing levels of specialization characterize the work, leading almost universally to decreasing accessibility. An alternative to this depth-first process is the less developed and organizationally complex breadth-first approach where interdisciplinary collaboration and the cross-fertilization of various fields are receiving the attention for generating innovative results.

In fields such as cosmology, where we only have one universe available, and the setup of the experiments is not under our control, it is possible to observe extremely high-energy phenomena that have deep connections to the theories of particle physics that would have possibly needed machines with energies that made them too large and expensive to be built on earth.

Comparative studies that leverage the capability of open access publications, and of the elaboration of big data streams and of the unstructured text of the articles themselves, can highlight statistical confirmations or anomalies in studies through the years, across several research institutions and the work of many scientists. This metaknowledge can lead to valuable understanding of reproducibility, effectiveness, and promising areas of research in order to better allocate attention and resources.

There is an increasing epistemological attention to the structure of theories, making sure that the excessive generative power of certain groups of theories is rightly questioned. String theory, a family of theories of particle physics, is able to pull out of its epistemological hat a theory that corresponds to any experimental result, given that it encompasses 10 to the 500 theories approximately (billions of billions of billions… of theories) with a questionable application of the sequence of theory, prediction, and verification.

The human element of the structure of scientific theories and the way they evolve is being understood more deeply. The capacity of embracing new, risky areas of enquiry is naturally more likely at the beginning of the career of a scientist. The increasing lifespan of prestigious leaders of academic fields must go hand-in-hand with a scrutiny about how they retain their nimbleness and risk-taking as the resources are allocated to varied approaches, and to give voice to new entries and new ideas in their respective fields.

EXPONENTIAL CHANGE

Dynamic systems

All phenomena are dynamic and while we tend to try to analyze and formalize them through an opposite process, it is the static abstraction that is the further representation of the reality. Of course, as the struggles in the evolution of the methods of science have shown, reality is not always readily decipherable. Nor it is conveniently laid out in front of us in intuitive packages conforming to our common sense. Many of our intuitions about the rules governing natural phenomena turn out to be wrong. This can happen with a series of experiments that, once the physical laws behind the phenomena are well understood, anybody can carry out. At that point, with accessible explanations that can be illustrated with immediacy, there is no excuse for being ignorant about the nature of reality.

A simple example of this is Newton's first law that says that "all objects maintain their state of motion in absence of an external force". Our everyday experience is that a car put in motion will actually stop, if you don't push on the gas pedal. But now we have a clear understanding of the role of attrition, and that the deceleration is due to the engine, the terrain, and the air in front of the car. Taking away all sources of attrition, the car will go on forever.

The consequences of dynamic change are all around us, in

the ebbs and flows of water, rivers, oceans, and rain. In the growth of vegetation, trees, and forests, or the advancing of deserts and the changes of seasons. But even if we have plenty of experience with dynamic change, our intuition can be more misleading than ever, with respect of the raw power behind its abstract mathematical nature, unencumbered and unrestrained by the constraints of a natural, physical environment.

Exponential change is such a dynamic environment. We can prepare for it, but its blunt force will still surprise us, often confounding even experts and certainly taking laypersons aback by how powerfully it can reshape the landscape of our reality.

Meaningful sequences

The simplest examples of exponential change, the doubling of a quantity, for example, in a certain amount of time, if the starting point is the unit of 1, can look fairly harmless or even disappointing at the beginning.

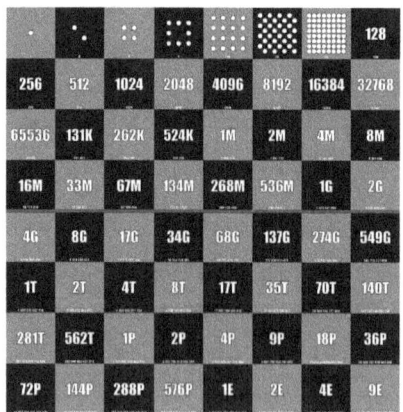

Figure 2: Doubling of grains of wheat on a chessboard

1, 2, 4, 8, 16, 32, 64, 128, 256, 512, 1024, 2048, 4096, ... is a sequence familiar to anybody who has had any interest at all in numbers. Rattling of the sequence in your head could have been a harmless childhood exercise.

There is a corresponding sequence, before the unit, that you can look at, potentially unassuming to even a greater degree: 0.01, 0.02, 0.04, 0.08, 0.16, 0.32, 0.64, and then 1.28. The interesting surprise of this sequence, nothing magical about it, is what came before: 0.00015625, 0.0003125, 0.00625, 0.00125, 0.0025, 0.005, 0.01.

Why are these three sequences interesting and why are they representative of the nature of exponential change?

Imagine that you are looking at the world around you and, trying to decipher it, predict what a given phenomenon will amount to, you collect data about it. This collection will not be as neat and clear as the sequences above. There will be a lot of noise in it. Errors of measure, mistakes made during the process or the planning of the measurements, other phenomena intruding and confounding your attempts to a clear understanding, and so on.

The noise of a natural environment where, before even being able to clearly train your ears to it, you want to discern a pattern that is possibly new, something that nobody else tried to listen to before.

Noisy signals

Is there signal in the noise? It is very likely that while you try to answer the question, there will be other opinions around. And per definition, they will be different from yours, not really aligned, or even maybe in opposition. If you are in a research environment and are competing for grants, or you are in an industry and the product that you are trying to engineer or the service that you are promoting among users rightly distracted by a huge offering of alternative options, in any case you'll be confused by the resistance to your original theory which the signals may support. You need to be strong in your opinions, you even have to have faith in what you want to show, an unreasonable conviction that you are right while everybody else is telling you that you are wrong. Or even that what you are looking for is non-existent, or impossible.

This is the realm of the third sequence, leading up to 0.01 (or 1% of the unit). The area where even experts will be against you. It needs a keen eye and ear, concretely or abstractly, to understand that in the presence of the distractions of a noisy natural world, there is indeed something brewing. Doubling calmly, without anybody else noticing it but you, and after several doublings arriving at the threshold of 1% from the goal of the unit.

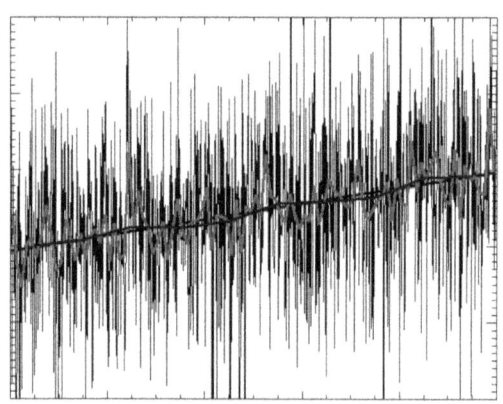

Figure 3: Interpolation of a signal from a noisy phenomenon.

After you arrive to the 1%, those experts who still don't believe you should be stripped of their label ignominiously. Because from then onwards it should be clear not only to you but to anybody who pays even a passing attention that it is now just a matter of time. In merely seven successive doublings you will have arrived at the unit.

The Human Genome Project's exponential growth

Turning this description from abstract sequences of numbers into a real example, we can look at what happened in the massive Human Genome Project in the US. Started in 1985, the duration of the project was originally set for 15 years. As with any scientific project, it was not completely clear how all the hurdles would be solved, and what approach would be the winning one. After seven years into the project only 1% of the goal had been achieved! Many at the time were loudly demanding that the project should be suspended or even abandoned: look, already halfway through and it is only at 1% to the objective! Those more careful, or the experts in exponential dynamics, though, were able to see that all was good. Having reached 1% doubling the amount of base pairs decoded each year in the previous seven years, in additional seven years of doublings the project would reach its goal of 100%, of decoding the entire human genome.

There are many phenomena that are subject to exponential growth: populations and nuclear chain reactions, to name a couple of examples. Populations grow exponentially because, as long as on average a couple has more than two offspring generation after generation, the increase will be cumulative: those offspring will have more children too. Nuclear chain reactions also occur when the fissile material, uranium for example, produces among its fission products neutrons that before exiting the volume of material make another uranium atom break apart, producing other neutrons, and so on.

Most importantly for the theme of this book, the power of computing and information systems is also growing exponentially, and has been doing so for over 50 years. However, there is no natural law behind this dynamic, no biological or physical necessity. It is an engineering project that ended up being called, by the name of the person who first formulated it, Moore's Law.

Moore's Law

Gordon Moore was working on the newly invented integrated circuit, at the beginning of the '60s. He was in a noisy environment, in terms of one having to concentrate on the features of a novel phenomenon in presence of many others going on at the same time. Practical computers have been around for a couple of decades, more or less, becoming more and more powerful, but at a rate that was rather slow, if looked at linearly.

Different approaches have been tried to make them capable of storing more information for the calculations, and executing them more rapidly. The vacuum tubes, magnetic core memories, and other components of what at the time the media liked to call "electronic brains" were cumbersome, prone to high failure rates, and needed scores of specialized personnel to take care of them, to make sure that they would work. The cost of computers was in the millions of dollars, and only national research programs, or very large corporations, could afford them.

The invention of the transistor, to be used as a basic component for calculation, promised much more reliable and cheaper production, assembly, and running of computers. Transistors could be packaged together with other components to create a useful unit of calculation called the integrated circuit. Not only that, but given their nature, it was possible to forecast the development of next generation components that were smaller, faster, and cheaper than previous ones.

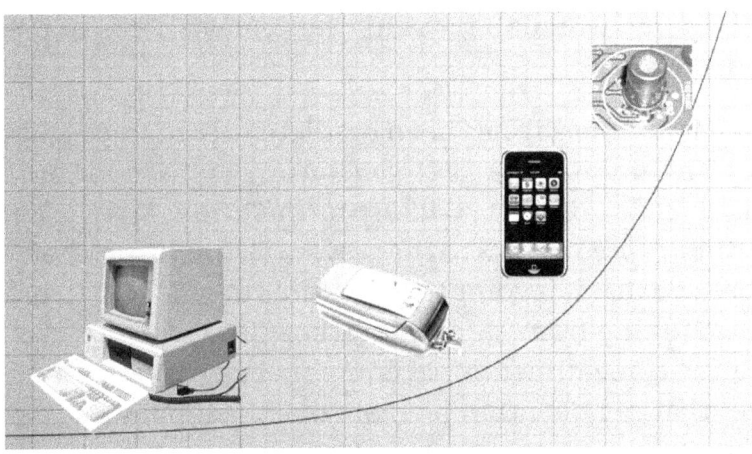

Figure 4: Computer evolution propelled by Moore's law.

Gordon Moore was able to observe what the current capabilities of the production processes were, and the increase of these capabilities in the course of a few years. Based on only a handful of data points, plotted on a piece of graph paper that after 50 years still survives, he boldly formulated a prediction that the number of transistors that could be accommodated on a given integrated circuit would double every year. A bit later he corrected the prediction to two years, and then finally settled on 18 months, which is the value currently accepted and used.

Based on how few data points were available to him this prediction was quite bold, maybe even reckless. However, with the benefit of hindsight, it would appear that this courageous ambition is what was really needed. Because what happened is that, spurred by curiosity, the desire to excel, and basic economic competition, more and more groups of engineers set out to create more powerful integrated circuits. Together with all the supporting systems that were needed, they weaved together an entire industry. At the beginning this process was driven by the individual capabilities of these groups, and what they were able to offer on the market. However, later on, Moore's Law became itself a driving force, a kind of self-fulfilling prophecy as well as a guide-post against which to measure the achievements of the various groups.

Many times it has been predicted that Moore's Law would fail to hold up in the next generation, and sooner or later it is bound to do so in its strictest formulation. More generally extending its predictions to the power of computing, there is reason to believe that it is going to be possible to still hold it up for a long time. Moving from silicon to other substrates for the circuits; creating three-dimensional components; moving from architectures that see quantum phenomena as a hindrance to ones that fully exploit them... there are many approaches to overcome eventual roadblocks that lie ahead in proving this law right, the same way others have been overcome in the past fifty years.

It is important to note that the spreading of knowledge is at the basis of Moore's Law. No single group working in secret could hope to be the one that will indeed be able to solve the problems that pop up along the road in the next generation of solutions, or the one after that and so on. Only the collaboration of many groups makes this possible. It is enough for one of them to achieve a breakthrough, discovering the solution necessary. All the others will leverage that, through licensing agreements that incorporate the solution in the next generation fabrication plants churning out integrated circuits, that are today produced by the billions each year.

The complex interlocking ecosystem of industrial infrastructure needed to maintain the pace of evolution in computing is not only in the production of the integrated circuits themselves. Similarly have to evolve the manufacturing tools that create the circuits, the software systems that allow to design them, and the financial support to make the investments possible for the plants, raw materials, refinement, and very importantly, human capital.

Whatever the fundamental physical limits for the growth of computation are, measured by the generalized Moore's Law, they lie far in the future. The progress that we've seen in the past 50 years of increase in the power of computing is going to be vastly eclipsed by that of the next 50 years. Actually, it is going to be exceeded, by the very nature of exponential growth, within the next couple of years. And then again, the next couple of years, and so on.

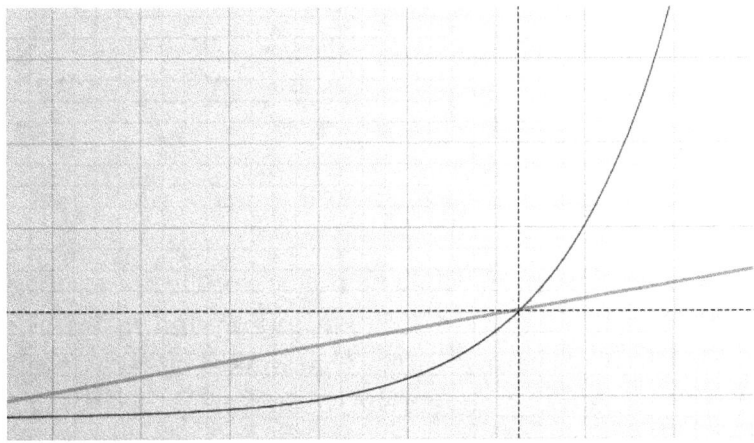

Figure 5: Linear progressions can overwhelm exponential trends initially.

The power of doublings

It doesn't matter, of course, in terms of how rapidly an exponential sequence develops. No need for it to double in a year to be exponential. These are just arbitrary units, and any cumulative change where the resulting quantity increases by a given amount expressed in the result itself will do. If you have a quantity of 100 and it increases by 10, you'll have 110, then 120, 130, etc. This is linear growth. But if you have a quantity that is increasing by 10%, then you'll have 110, 121, 133, and so on. That little difference, which doesn't appear to be very significant at the start, is all that matters. That is exponential growth.

There are many ways to express this power, and how surprising it is for those accustomed to think linearly.

Look for example at the sum of this sequence: 1, 2, 4. The sum 7 = 1+2+4 is the total amount in the entire sequence. And the next step in it is 8, larger than the total of all the steps preceding it. This is true for all exponential growth. In the next period of doubling in computing in merely 18 months, thanks to Moore's Law, there will be more transistors and integrated circuits created (and computers from them and calculations carried out through them) than in the entire history of computing for the past fifty years or more!

When is it too late?

Another example illustrating the power of exponentials is to look at a closed system, for example a pond sustaining a population of frogs. If there are algae that make the lake uninhabitable for the frogs, and as it covers the surface of the lake in ever greater extent, from just a fraction, to one percent, doubling every week, how long do the frogs have to live, once the algae covers half of the pond? By now hopefully the answer is clear: only one week, as during the next doubling the lake will be entirely covered by algae! Even more alarmingly, perhaps, already at 1% there is less than two months' time left for the frogs to flee to another pond, or to find a way to stop the algae from expanding.

Our unique position is to be able to see what is happening to the pond, contrary to frogs. And this capability of data collection, analysis, and foresight gives us great responsibility in understanding if the pond is OK or not. Taking active action managing the pond, countering the algae, can't and won't be done by others, but we can do it.

The various examples of exponential change we find in

nature feed themselves, but seldom assemble into interacting chains that feed each other. Our technological civilization on the other hand is full of these self-reinforcing chains that keep the acceleration of exponential change going.

Ray Kurzweil's project

The inventor, author, and co-founder of Singularity University Ray Kurzweil has been collecting data about exponential phenomena for decades. It is not enough, in fact, to be able to recognize what is going on. The explosive nature of exponentials is such that timing is crucial if you want to be able to ride their wave instead of being swept aside by it.

Jump on the wave too soon, and those who say otherwise will have an easy time deflating your enthusiasm or that of your financial backers, because the upswing from the sequence trailing the hypothetical unit in our example sequences will not happen. See it too late, and it will already be in full power when you'll want to climb it, making the endeavor too expensive, difficult, or even impossible, as others will have crowded the crests already.

From flatbed scanners to optical character recognition, from musical synthesis to speech synthesis, or handheld systems for the blind, all of Kurzweil's inventions take advantage of a keen understanding of the right timing. When to accelerate research and development, so that by the time supporting hardware systems are available at the right price and the right level of integration, all the other components of software, user interface, development systems, and the entire supporting ecosystem are ready as well.

At the Santa Fe Institute based on research by Bela Nagy there is a full database of exponential phenomena that is accessible to be studied and expanded upon further.

Kurzweil also recognized that with the interconnected and

intercommunicating systems of human knowledge that do not grow in isolation but reinforce each other, there are exponentials feeding on exponentials. He called the resulting effect the Law of Accelerating Returns. This is contrary to the acquired wisdom of classical economics where it is assumed that in order to achieve a given increase of economic of output there needs to be a progressively higher amount of available input of capital, called the Law of Diminishing Returns.

Just as with Moore's Law, the Law of Accelerating Returns formulated by Kurzweil is a self-fulfilling prophecy, sustained by the open communications, and competing groups aiming to achieve success and excellence in their research and industrial production endeavors. It is definitely possible to break either of these laws. If you stop believing in the law of universal gravitation and jump off the fifth story of a building, you can do it a thousand times, and you'll never stop dropping like a stone, most probably to your death. But if we were to give up trying to make better circuits, or if we decided that it wasn't worth our effort to make better solar panels, better batteries, and so on, as long as everybody stopped, those circuits, panels, and batteries won't happen.

Kurzweil is, as of this writing, a Director of Engineering at Google, by his own admission the first ever job he has had. Using the resources made available by the company, he is applying his skills to make natural language interaction with the computers possible, and the next wave of user interaction, making computers even easier to use and able to better serve our needs.

Connecting S-curves

A frequent criticism of Kurzweil's analysis and predictions is based on a misunderstanding of what constitutes the exponential that he is talking about. The critics highlight the fact that what appears to be an exponential in reality is the first half of an S-curve or a logistics curve. It appears exponential initially, as the benefits of a given technology are exploited. However, after a while it plateaus out because it becomes harder and harder to squeeze additional benefits from the same technology. It becomes exhausted, and the belief in the power of technology of those who preach unending exponentials falsified.

In reality this is true: each individual technology as it runs its course is unable to give more than its natural limit. As it approximates that limit, it becomes fruitless to insist to want to get more out of it, both from the point of view of engineering as well as from that of economics and of return on investment. And that is why new groups with new ideas will try to achieve the desired output through a different approach. Smart people will see the point in time when the current generation of technologies gets exhausted, and work in parallel with the groups leading at the time, to find a new technology that will deliver the objective at scale, better than before. The cycle in a few exponential doublings will repeat itself, and a third generation of solutions will be needed, and so on.

The cumulative effect of these different S-curves, smoothing out each other's endings, and more or less seamlessly interlocking in a chain of invention, innovation, and industrial deployment, is drawing the exponential that Kurzweil points to in his analyses.

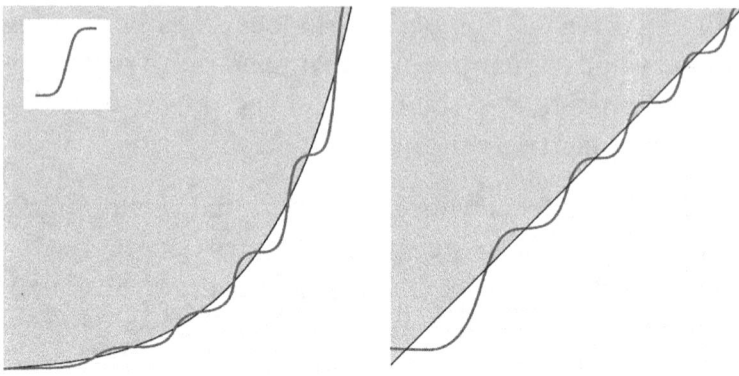

Figure 6 : A sequence of S-curves designs the exponential trend.

Taking computing, for example, there have been many generations of computing technologies, each leading in its own time, which have been pushed to their limits, and superseded by the next one, better, cheaper, and faster to generate the desired output: calculations. Mechanical relays, vacuum tubes, transistors, and integrated circuits decade after decade enabled the construction of the world's fastest and most powerful computers. The companies employing them were the leaders of their time, pushing the limits of the technologies, and were supplanted by new models, based on the new technologies in a few doublings of performance.

Another example where these years we are witnessing a fundamental switch is in permanent memory storage. The larger and larger amounts of data that our computers need to persistently record, so that when electricity is turned off, and the computer wakes up later on, the data can be retrieved without having to start over from scratch. From punch cards, to magnetic core memory, to magnetic tape, to spinning hard disks, now we are on the verge of moving storage for next generation needs to solid state support (flash storage), which is going to be able to memorize orders of magnitude more, accessed much faster and reliably, and more affordably than any generation of device previously.

Exponentials everywhere

Many technologies can be seen through the lens of this exponential interpretation of accelerating change. The doubling periods can be different, of course, than the 18 months that Moore's Law accustomed us to rely on.

In solar energy we talk about Swanson's law, which represents the decrease in price per watt of a photovoltaic panel. Starting in 1974 with the creation of the first of such devices, that cost over $70 per watt, today we are at $0.30 per watt, and the decrease in price continues. This decrease comes from economies of scale, from better understanding of manufacturing processes, and from the birth of an ecosystem of financing, deployment and servicing of the modules, as well as from new basic approaches of materials and construction methods that increase substantially the efficiency of a given module as it transforms sunlight into electricity.

There is a doubling of the storage capacity of our batteries. This doubling is a more sedate (and infuriating if you feel you spend too much time charging your various power-hungry devices) ten-year period. Depending on metallurgy, chemistry, and manufacturing processes, it is not unimaginable that, in an illustration of Kurzweil's Law of Accelerating Returns, the industry would find a way to speed up the doublings, by adopting a radically new approach and making applications practical that would have been impossible before.

No magic 100%

The goals that research programs such as the Human Genome Project set themselves are often somewhat arbitrary. They represent a useful goalpost, but not the end of the development of processes, their refinement, and certainly not the desire for knowledge and for the capacity to acquiring it faster and cheaper. After decoding the genome of a single individual there is the task of doing the same for another seven billion of them. After the human genome there is the genome of other animals, or bacteria in the oceans, or the bacteria that symbiotically live on and in all of us, constituting what is called our microbiome.

The capacity of decoding the human genome did not stop at the rate of one per three billion dollars in fifteen years. We certainly would not have taken much advantage of that. In the fifteen years from that first success, the technologies that have been invented, perfected, deployed and substituted by better ones again, allowed an astonishing progress: today it is possible to decode an entire human genome for about two thousand dollars in a couple of weeks. But progress is not stopping there either, and it is possible to forecast the availability within the next ten years of technologies that will enable the decoding of a genome for less than ten cents in a fraction of a second. It is worth thinking about the transformations that this kind of change is going to bring in the world of healthcare, insurance, privacy and more.

The takeaway is that there is nothing magical about a given threshold that we end up calling a unit or 100%, and that the power of invention and implementation that drove technologies to achieve that doesn't stop, but keeps going on, delivering increases in the desired output, at lower costs, faster speeds.

ARTIFICIAL INTELLIGENCE

The Nature of Intelligence

The meat of this book could be a philosophical analysis of what is intelligence, and whether it is possible to describe it in scientific terms, or, rather, if it represents an essence that is irreducible and irreproducible.

Indeed, this has been the occupation, in various forms, of philosophers for thousands of years. Including the nature of truth, beauty, right and wrong, morality, and ethics and esthetics, philosophy had the run of the field, unencumbered by practical considerations. To the contrary, the division of knowledge into two fields of abstract understanding and of that with practical consequences, with this second being frowned upon and seen as inferior by the followers of the first, has been a mainstay of Western philosophy since Aristoteles.

There are several assumptions in this book, and the fact that intelligence can be understood, analyzed, and reproduced is one of its fundamental ones. In more recent books by philosophers who don't disdain being understood by many, there are wonderful arguments to why this should be the case.

In general, for the purpose of this book we will define intelligence as the capacity of matter to organize in a way that allows it to seek solutions to goals through plotting a path of action towards them, and organizing both abstract and concrete resources to achieve them.

The human brain is a lump of matter that is endowed with a certain degree of intelligence. And on purpose I use that expression instead of "mind". We will avoid the pitfall of dualism that has bogged down philosophy trying to understand a mind that just inhabits the brain, dragging Descartes into arguments about homunculi and fruitlessly searching for the connecting tissue through which the mind is attached to the brain. Under this assumption the brain expresses the mind, and the mind is what the brain does.

The Mechanical Turk

During the 1700s a fascinating contraption toured the courts of Europe's kingdoms. Designed by the Hungarian Wolfgang von Kempelen, it was a large box on top of which sat a wooden doll; today we would call it a robot. It played chess, and beat anybody attempting a game with it, unerringly, with mechanical precision. Von Kempelen's claim that he built an automaton capable of acting with intelligence was unraveled when it became apparent that the box housed a midget. Indeed good at chess, the midget pretended to be the intelligence behind the robot. It actually was the homunculus of an artificial setup for mechanical intelligence. It constituted an artificial artificial intelligence.

Turing tests for humans

When the first electrical computers were born, in the '40s, Alan Turing formulated a novel test for the intelligence of machines. The Turing test, as it is called today, asserts that there is every reason to believe that a machine is intelligent, if it is deemed to be intelligent through its actions by a team of human judges. During the test the setup hides the machine from the humans, and mixes it and its output with other humans that during the test can pretend to be machines wanting the judges to believe them or the humans they are. Turing called it "the imitation game", probably believing it to be less fundamental than the fascination with which it is seen now. Too frequently and somewhat bombastically one or another of the media sources announces that the Turing test has been passed. Universally, when the transcripts of the dialog typed out between the winning machine and its human judges becomes available, it appears that the machine, or rather its programmers, hid behind pretend jokes, changing of the subject, and other trivia.

Chess playing experts

In 1996 the then World Champion of chess, Gerry Kasparov, was beaten by a machine called Deep Blue, specially designed and built by IBM for this purpose. The machine could definitely not attempt to pass the Turing test, in its imitation game based on written dialog encompassing general subjects. However, with its hardware and software specialized in analyzing and discarding millions of moves in the tree of possibilities of the game of chess, until it found what it decided was the best one to make in a given configuration, it nonetheless faked to be intelligent in the game perfectly enough to beat and even infuriate Kasparov.

If the Turing test, the imitation game, consists of convincing a group of humans that a machine can have a human-like dialog, following the rules of syntax and semantics, then the Kasparov test, the chess game, consists of convincing a chess player that he lost to a machine that plays following the rules of chess. And relevantly enough, in this case there is actually no difference between the machine that pretends to know chess and one that does know chess.

During the various stages of the tournament between Kasparov and Deep Blue, the engineers from IBM tweaked the algorithms of the machine, and the Russian champion strongly protested. The machine should not be able to learn during its play? The tweaks that constituted an increase in its smarts, and possibly contributed to its winning, were admitted by the judges of the tournament.

Expert systems and restricted artificial intelligence

Running on specialized hardware like IBM's Deep Blue, or more general computer architectures, even personal computers, systems that exhibit the decision-making power of a human in a given specialized field are called expert systems. The field of artificial intelligence (AI) was dominated by the approach of expert systems in the '80s. In fields as diverse as medical diagnostics or financial planning, the knowledge of experts in a given area was encoded in rules that were put in motion by inference engines capable of applying them according to the data provided, in order to generate a recommendation for a course of action: what possible illness the symptoms suggested, or which loan was most appropriate for a given financial situation.

These expert systems were relatively successful, and are still employed in various fields, but did not represent an attempt to create a general model of intelligence, and could not be the stepping stone on the path towards the general artificial intelligence that would be able to be an expert in any field whatsoever.

Hopes and disillusions

Many of the original practitioners of AI have held the belief that they could quickly build computers that exhibited higher capacities for thinking, creativity and problem solving. They were taking advantage of the stimulating environments of academia, from MIT to Stanford and elsewhere, to establish laboratories in the '60s studying what was possible. And many of them left academia to raise funding from industry or venture capital in the hopes of creating scalable and sustainable innovation through the application of what they learned in the labs.

Most of the claims that were made, even taking into account the constant development of the hardware that was available to be at the basis of the software systems for AI, went unfulfilled. Or at least they did not achieve the scales that the funders needed to justify their continued investments. By the '80s, what was called "AI winter" descended, and it looked like the field would not change the world as deeply as originally thought.

This is a common effect of misunderstanding exponentials. Overexcitement by the outsiders of their feeble understanding of the underlying principles, coupled with the eagerness of experts to deliver solid results, looks at a few data points with a linear interpolation. But the linear growth at the beginning of an exponential is actually higher! So those who underestimate the power of the exponential further on are also bound to make the mistake of overestimating it at the beginning.

The role of learning

When computers were born, their architecture initially was not that of what we today recognize as a computer. It more resembled a specialized tool that could be used only for a given purpose, rather than that of the universally adaptable instrument we use daily today. The hardware was designed to be optimized and literally wired for that single task, and it was not feasible to rearrange it to do anything else.

Only after a while with the stored memory computer, implemented with the Von Neumann architecture, which did not distinguish between numbers representing data and numbers representing instructions, was it possible to talk about a universal computer. Even then an additional development was necessary to complete the concept of programming the computer, and of representing the programs in higher level formalisms, abstract languages that could then be translated and compiled into machine language, the steps directly executed by the computer.

Writing these programs was a new art at the middle of the 20th century, and even if certain basic components, like branching and loops, had been conceptualized already before they became practical elements of a running program, more sophisticated tools needed to be developed to be able to juggle larger and larger programs, making sure that they would be able to execute without problems.

How the programs could be written, and if the programs themselves could be looked at as data, with other parts of the program rewriting them as needed eventually, changing the behavior of the main program being executed, was something that Turing already considered, and likened to the role of learning in humans.

If we can build a program that plays chess, another one that makes medical diagnoses, or financial recommendations, can we build programs that are good at all of those things and others? Can we build a program that, running on a sophisticated and powerful computer, can be put in front of a problem, any problem, and find a way to analyze it, to garner the resources needed, and to solve it? Can the program look at its results and decide if they were optimal, or if, with additional data available, there would be now a better way of achieving success? To learn and solve problems, in a completely universal manner where its programming is not fixed but fluidly adapts to the needs represented by the environment? That is what we call Artificial General Intelligence.

ARTIFICIAL GENERAL INTELLIGENCE

The assumption of this book is that it is indeed possible to build what is needed for a behavior to emerge that can analyze and solve an arbitrary set of problems. Build the hardware that is powerful enough, in terms of speed of execution, and in terms of its capacity to store and access the memory needed, as well as in practical terms, that this hardware would be manageable, could be powered with a feasible amount of energy, and it would be possible to build it with the resources that we have or we will have available when we'll know how to build it. And also that it is possible to build the software, the set of programs that run on the hardware, that are able to acquire the data to recognize the problem at hand, to access and dynamically use the knowledge base to derive from the possible approaches to attack the problem, and which is flexible enough to combine these approaches in novel manners, or to create new approaches altogether, in order to optimally solve the problem with the resources and the data available.

The Turing test becoming moot

The original prediction by Turing made at the middle of the 20th century that the Turing test would be passed within 50 years did not come true. We do have a constant dialog with our machines, and actually, with the disappearance of keyboards into smaller and smaller touch screens, or the disappearance of the computers themselves into the environment, conversational interfaces are a natural way for human-computer interaction. However, we are in no illusion of having a dialog with a human when being addressed by a machine and establishing a conversation with it.

In some sense this is to be expected. The goal of faking to be human is meaningful only from a Hollywood point of view, but not necessarily useful beyond a certain point. A robot will benefit from a humanoid form, from being bipedal for example, or from having hands, as it will be able to better navigate a human environment full of steps, stairs, doors, and handles. But once it is able to do so, additional efforts to look like a human are a good investment only if it is proven that, for example, the psychological reaction of humans is better towards a human-looking robot rather than a robot-looking robot.

Similarly, a conversation that is aimed at a useful outcome, of for example booking an airplane ticket taking into account dozens of constraints and millions of possible combinations of airlines, flight times, connections and seating options, will not benefit from the addition of human centered quirks that make it more likely for the machine to be seen as human instead. An occasional cough, a sprinkle of jokes, or an aside about a remark that can occur in the conversation will be supported by the dialog system only if it results in more tickets being sold faster, and with a higher degree of satisfaction on the side of the human caller.

The value derived to the human counterpart from the interaction with the robot, or the conversation with the machine, is a goal in itself. There is a yearly contest financed with monetary prizes that keeps the spirit of the Turing test alive. The chatbots that participate employ a full gamut of tricks to throw the human judges off and into believing that they are human. Many of them are also accessible through an online web interface for anybody to experience a conversation with them. However, as also illustrated by the relatively modest amount of investment going into the contests yearly, there is a general consensus that the direction of really useful research is elsewhere.

Unavoidable anthropomorphizing

Together with several other threads of philosophy, unsettled for thousands of years, the difference between substance and emulation is reverberating in the Turing test. Turing's conclusion is very practical, though: if there is no statistically meaningful difference in the output and its effects, then we have no reason to assume that there is difference in the substance.

From an epistemological point of view, of course, this is not at all true. We can have a system appear to be identical to another that it is emulating for thousands of different combinations of input, and then, suddenly, generate an unexpected output for a given set of inputs, totally different from what the original would. This has been exploited in numerous works of fiction or Hollywood movies, where the initial assumptions become dramatically falsified.

Figure 7: The intentions and objectives of robots will be different from those of humans.

The way human perception works, it is natural and unavoidable to project human qualities and characteristics on non-human objects or beings. From our childhood toys to the behavior of dogs and cats, or the way we describe the behavior of appliances that do not carry out our instructions the way we would want them to, the temptation of endowing each with human-like features is irresistible. Intention, desires, will and free will, emotions, empathy, and many others are endowed on them, with the consequence that their behavior is assumed to include a wider set of options exhibited by human actors. This is a useful shortcut that allows to succinctly say that a television set "goes to sleep" as its timer is set appropriately, and an unlimited list of other convenient turns of phrase. Nobody would then generalize and attribute to a television set broader human-like features and behaviors.

One of the questions that is going to be crucial and discussed more in detail further on is when will this distinction stop being meaningful? Until then, it is going to be useful to keep in mind that the expressions attributed to complex systems in describing their behavior are part of a metaphor that does not in itself imply equality.

Predictions for AGI

Most of the people who are working professionally in the field of artificial intelligence see no theoretical barrier to creating an Artificial General Intelligence (AGI) as described above. There is some disagreement on the fundamental nature of the result, and a fairly widely distributed set of forecasts about the time when creating an AGI will be achieved.

There is an informal survey that polls AI experts and plots their answers on a time scale for a successful AGI implementation. From as far out as a hundred years or more, after then end of the 21st century, responses in later polls have started to cluster around the middle of this century, with a dispersion in the replies that is narrowing too.

Architectures for AGIs

The two main routes towards AGIs consist of understanding and emulating how the brain works, and in reimplementing its flexible problem-solving capabilities through different means.

Neural networks and what are now called deep learning algorithms allow a system to make decisions around complex inputs and possible outputs, using a feedback mechanism that does not require the specific rules that govern the decisions to be made explicit. Simply running the system through a simulated scenario where the positive and negative outcomes at every step are clearly noted, and generating variations in the decisions in order to allow the system to try out a wide variety of options to pick from based on the feedback received, given enough time and computing resources, will generate astonishingly well-performing results.

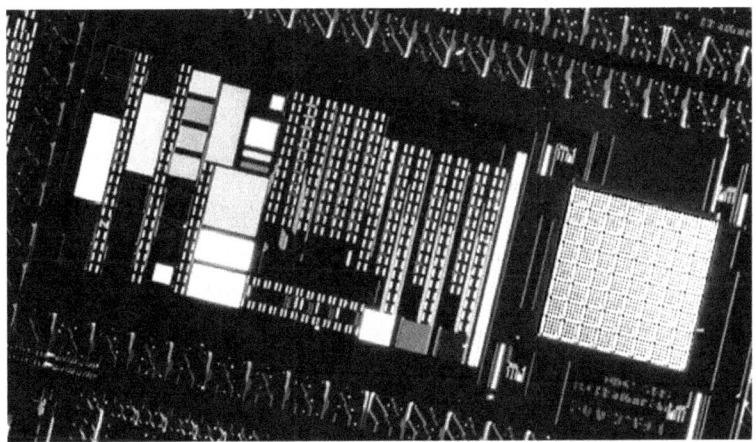

Figure 8: Quantum architectures promise a radical increase in computing performance.

Applying these deep learning approaches to dozens of

different video games from the '80s, it is now possible to evolve a system that not only plays the game well, but plays it better than any human can. Originally these games were running on their own hardware, and in isolated cabinets, coin-operated, within amusement arcades. Today they are themselves living inside larger computers that are able to emulate their hardware with complete precision, as well as the software program running on them. Later games, again learned by these algorithms with superhuman performance, are in first generation consoles. In either case, it can be argued that the full set of games represent different problems in a universe of video games, and that in this sense the capability of the deep learning approach to master them with very little or no input about their goals, rules, input mechanisms and so on, is, within the constraints of that given universe, the behavior of an AGI.

AGI hardware

We are approaching the limits of the traditional silicon based transistor, and new steps in a generalized Moore's law will have to be taken through different substrates and different hardware architectures.

Already next generation chips are designed using CAD, computer aided design systems, which are in turn powered by current generation chips and software, effectively co-designing not only hardware with software, but also more powerful computers with less powerful ones. It is natural and likely for AGIs, while less fully formed, to already participate in the process.

Computronium and Jupiter Brains

The theoretical extreme of the increase in processing power as we organize matter to calculate is called computronium. Very simply, regardless of what atoms it is made of, or how it is structured, it represents the densest possible form of matter for calculations. Consequently the only way for computronium-based systems to increase their power is to increase their mass.

The very powerful AGIs made of computronium the size of a giant gaseous planet are called Jupiter Brains. Still possibly hungry for more computation and more matter to convert to it, they scout a solar system for other planets to eat.

An ontological argument for the speed of light to be an upper limit to signal propagation that no future development can overcome, related to the simulation argument described towards the end of this book, comes from it having the natural consequence of an upper size for Jupiter Brains. As the left side of it wants one thing, seeing something to eat over there for example, very simply there is no time to agree with the right side that may want to go the other way, before both do and the object physically breaks in two parts.

Self improvement

The objectives that a given system has to reach define its architecture, components, resources, way of working, and outputs. Depending on how complex the objective, the path to reach it can be direct and evident, or in itself naturally composed of intermediary steps. Some of these intermediary steps may be easy and uncontroversial, while others less evident, or clearly presenting alternative approaches. Selecting among the alternative approaches may depend on the previous results, or it could be the case that there is little reason to pick one instead of another beforehand. After the fact, it may be possible to establish that the option chosen was, if not the best, one of the better ones, or on the contrary inefficient.

The more flexible a goal-seeking system is in organizing itself in order to reach its objectives, the more explicitly it is going to be dedicating part of its resources to these types of considerations which are not about the goal, but the means, the tools and the methods to reach it. Meta-reasoning, reasoning about the reasoning: an opportunity to become better at the task by realizing what the best ways to reach it are, and using those, rather than alternative, inferior ones.

Most of the approaches to AGI incorporate learning algorithms that implicitly or explicitly allow the system to apply meta-reasoning. An AGI system consequently will get better, and will improve in time, achieving better performance at a given task, or being able to pursue more complex goals with a given amount of resources.

Intelligence explosion

A system that is tasked with reaching a complex goal, and has the capability of analyzing and improving on its own behavior in completing it, will take advantage of that capability. It will improve itself, in order to reach the goal faster, or with fewer resources. If we see the capacity of reaching that goal a given level of intelligence, then a better way of reaching the goal is a sign of a higher intelligence. The system gets smarter. However, this process doesn't stop by itself. It will, on the contrary, feed on itself in an exponential fashion.

A smarter system will be not only better in reaching its goals, but will also be smarter in analyzing the ways that the process can be improved. It will apply the results of this analysis to itself, and then start the cycle again. The process through which this iterative self-reflective improvement occurs is called intelligence explosion.

Self-awareness and introspection

The degree with which a system is able to perceive its environment and to derive useful decisions from it is called awareness, at least in the case of humans. And the similar degree with which the same process is applied to inner states and parameters, rather than those of the outside world, self-awareness, and the process of data acquisition is termed introspection.

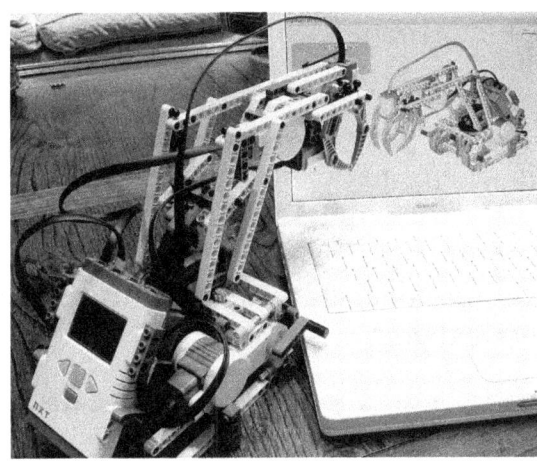

Figure 9: Self recognition leads to introspection and self awareness.

With the caveat of applying these terms loosely, during the intelligence explosion the AGI systems become more aware, more self-aware as their capability of introspection increases.

Open access to your self

During the ten-thousand-year history of our technological civilization (or a hundred thousand years if we want to be generous and start with the adoption of fire rather than that of agriculture), we struggled in giving a solid basis to the understanding of our own being. Only very recently we have begun to understand how the biological recipe of DNA gives rise to embryos and then individuals, and are barely scratching the surface of the complex interactions that the possibilities of our genetic options express as they interact with the environment, and with our learning.

By applying a metaphor of business models, we can say that humans saw themselves as a closed source proprietary system, with no user manual, no administrator's guide. We had to slowly reverse engineer all the components of our bodies (and the world around us), and indeed, it took as an understandably long time. (Hopefully nobody has taken out patents on the design of the Universe, and is ready to sue us for infringement!)

It is natural to assume that the AGI systems we will build are going to be judged by how well they perform. Consequently, since they will be able to perform better if they can improve themselves, those that do will be preferred. It will be then obvious to help them along, contrary to humans, by giving them access to their own source code, along with full instruction manuals on how to access and improve it.

It won't take ten thousand painful years for AGIs to realize the DNA they are made of, or the binary code, rather. They will be born aware, self-aware, and in full capability of acting on their introspective powers.

Slow takeoff

How will AGIs impact the world? According to most of those who study the field, once invented, it is not going to be possible to uninvent them, to put the genie back into the bottle. Only a universal planetwide relinquishment of the tool and its benefits would be able to stop AGIs from being used, deployed, and profoundly influencing the world. It is believed that the business benefits alone will be so dramatic that it is inconceivable that corporations would not take advantage of their superior capabilities of optimization and problem solving.

As the AGIs that are open source will be performing better than those that are proprietary, their availability will spread, and their benefit will accrue to the widest possible group that is able to take advantage of them.

Similarly to how the electronics industry through cross licensing deals spreads the benefits of a single group's invention until it is adopted universally, constituting a stepping stone to the next generation solutions, AGIs will spread innovation in business models, social organization, and impact the lives of individuals, transforming everything around them worldwide.

The school of thought that is called slow takeoff describes this process, fueled by the intelligence explosion, in terms of decades.

Rapid takeoff

The school of rapid takeoff says that you go to sleep, and when you wake up the world around you is unrecognizable.

Much of what is discussed in this book is a staple of science fiction stories. Some of it benefits from the reader's suspension of disbelief, and there are assumptions, often explicit, about how technological development will happen, and what is indeed possible theoretically or at a practical level.

The vision of a rapid takeoff, as described above, whatever its concrete form might be, of a transformation so fundamental as to encompass the whole world to make it radically different in a manner of hours, is squarely in the realm of those that stretch the imagination.

The capabilities of AGIs to corral resources to their goals, and the transformative power of their innovative solutions to their ends will certainly be unprecedented. How quickly will a self-improving AGI start using knowledge only accessible to it?

Scales of intelligence

Previously when describing the arbitrariness of a given goal's 100%, the subject matter was DNA, and biology. But it is probably clear that human level intelligence, to be exhibited at a certain point by the problem-solving capabilities of AGIs, is similarly arbitrary.

The intelligence explosion of self-improvement will pay little attention to supposed IQ values of 100 (the average, per definition, of any group of humans), 140, above which one is considered a genius, or 1000. It is going to be difficult to measure the intelligence of AGIs in traditional ways, based on the speed of solving certain problems not only of mathematics, but also of verbal dexterity. Speed, robustness, flexibility, and creativity will be the criteria to evaluate these new kinds of intelligences. Assuming that new scales for measuring IQ will be devised to include specific capabilities of AGIs, there is a possibility that compared to a human 100, on any of these new scales theirs could be on the thousands or millions.

It is not easy to imagine in what ways an AGI with an IQ of 1,000,000 would manifest itself. How it would decide to interact with humans? The analogy of our inability to usefully interact with ants, and the limits of our positive but constrained interactions for example with dogs can be a meaningful if alarming one.

Is super intelligence uncontrollable?

There are many nightmare scenarios that can and have been developed around the rising of AGIs, super intelligent machines, in novels and Hollywood cinema, but also recently in more formal scientific settings.

What are the boundaries of action for an AGI? How can we make sure that its impulse to optimize the resources it has available, or it can make available to itself, is kept in check?

If the power of AGIs is as great as it could appear from preliminary analysis, then making sure that their actions are positive for humankind is essential. The consistent, assured and reliable friendliness of artificial general intelligences to humans and humanity as a whole is an existential challenge not dissimilar in its impact to the one dinosaurs faced against their asteroid.

Can we make sure we'll be different? Can we engineer an ethical system that will be followed by AGIs as they develop goals that go beyond what they've been originally given? Is it conceivable to create boundaries and constraints that will bind their actions within certain limits?

In the domains of unknown, between known unknowns and unknown unknowns, the second is more dangerous if left that way, or if the state of unawareness about them persists. It is not per se a radical cause for alarm not to have exhaustive and reliable answers to the fundamental questions above. But it would be irresponsible and irremediably so to neglect investigating the questions, seeking answers, and to assure that engineering these capacities didn't go ahead without a deeper understanding of the consequences.

AGI getting out of the box

The safety requirements of certain technologies that are thought to be potentially very dangerous brought the development of effective containment protocols. The discovery of recombinant RNA technologies and the possibility of gene therapies was discussed in the '70s at the Asilomar Conference that adopted procedures we now know were effective: there hasn't been in the forty years since a biological accident that involved errors around these technologies.

Recently there has been an Asilomar Conference on Artificial Intelligence, explicitly discussing what are possible containment procedures around advanced AI, and AGIs, as well as their dangers and impacts. Keeping an AGI in the box, so to speak, disconnected from the internet, limiting its computing resources, and making sure that it can't commandeer other resources to its availability than those initially allocated.

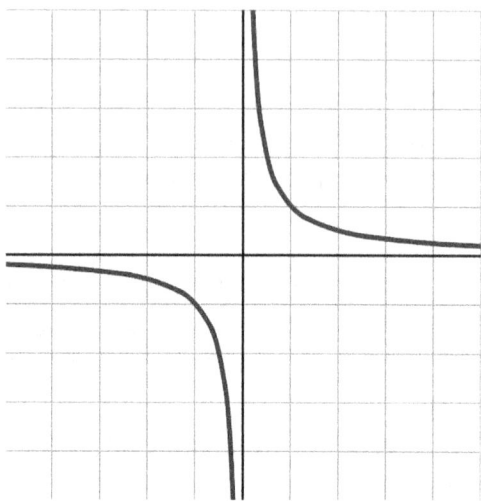

Figure 10: Discontinuity over a mathematical function.

Many believe that it is not possible to avoid the AGI getting out of the box. With reasoning, interaction, conversations, arguments, tricks, pleading, applying rhetoric, and recourse to ethics or moral arguments, it will do everything to finally successfully persuade its keepers and guardians to allow it to get free.

Singularities

In mathematics we speak about a singularity at the point when a function loses meaning. There are simple examples of this, like the function $y = 1/x$ which has a singularity at the point $x = 0$. As you approach zero, the value of the function, y, tends to infinity, and at zero it does not really become infinity, but undefined.

The problem in the example is not infinity itself. Mathematics has been extended to deal with infinity, actually different varieties of infinities, and not to shy away from their existence, but to usefully manipulate them. The issue of the example is the inconsistency, the fact that there is no given way of handling the point of singularity.

There are many types of mathematical singularities, and mathematicians have become very well versed in dealing with them. A common way of taking away a singularity is to assign the value of the function at it in a manner that makes it smoothly connect to the other parts, for example.

In physics the term singularity is applied to situations where the values of certain parameters would go to infinity, and the laws describing the dynamic evolution of the system stop applying. A classic example of a physical singularity is a black hole, the final stage in the evolution of a certain class of stars. When stars that are massive enough lose their ability to generate energy through fusion reactions after exhausting the available material, they can become supernovae, shedding their outer layers in immense explosions. The remaining shrinking nucleus will become ever denser. It will crush the atoms constituting it, overcoming the resistance between protons and electrons, and end up in a state of condensed matter called neutronium, as they end up electrically neutral like the neutrons of the atomic nucleus, and we call them neutron stars.

But if their mass is larger than a given amount in a small enough radius, then their density will keep growing, and not stop at the neutronium stage. The gravitational force will be so strong that the escape velocity from these stars will increase to values that exceed the speed of light. According to the theory of relativity nothing can move faster than light, and these objects stop emitting it, but also will become a one way street. The surface around them where the escape velocity exceeds the speed of light is called the event horizon. Any object whose trajectory brings it within that surface will never be able to escape it.

When black holes were first theorized, and then observed —not directly of course, but because of the lack of a star or any meaningful radiation in the middle of an orbiting system with characteristics that should have one—the first impression was that nothing could be known about them. However, physicists soon started to ask themselves what would happen if black holes were rotating instead of being static, or what it would mean to apply the assumptions of quantum mechanics and its principle of uncertainty to particles around the event horizon. And quickly, instead of being seen as completely intractable objects, due to their containing a singularity, similarly to mathematicians with their singularities, physicists found ways of studying the nature of black holes, classifying them into families, predicting their future histories, and so on.

THE TECHNOLOGICAL SINGULARITY

The term Technological Singularity was introduced by Vernor Vinge at a conference organized by NASA in 1993, and it represents a moment in time when with the introduction of AGIs the possibility of useful predictions about the future stops. The intelligence explosion, and the arbitrarily complex tasks that AGIs can attempt, their vastly different ways of reasoning and organizing resources are, at first approximation, such an infinity in the field of forecasting as their own singularities in the fields of mathematics or physics.

And in fact, the same way as mathematicians and physicists have not been deterred by the dangers of infinities from studying and usefully handling the singularities in their fields, technologists have started to attempt to understand the types of technological singularity that we can model, and classify them and the AGIs constituting their active catalysts.

There is hope that, by applying resources and the right level of effort and smarts, when AGIs will appear, on one hand we will be able to seed them in a way that will have them behave in a friendly manner, building a world that is compatible with human life and aspirations. And on the other hand we will be also ready, adapted to live a fruitful life in a world that is profoundly transformed by their presence.

Kinds of minds

We are accustomed at looking at intelligence as a single unified phenomenon, experience and tool. *Homo sapiens sapiens* is alone on the planet with the capability to observe and analyze its own awareness, self-conscious state, and describe and communicate it in rich and nuanced manners. Being the sole species with a given characteristic is surprising. As if we were the only species with eyes, or the ability to perceive and interpret sound waves, hearing. It hasn't been always the case.

At certain points in time, different tool-making and fire-controlling species of evolved apes lives on the planet, sharing it, without necessarily being in contact. The last of one of these, *Homo sapiens neanderthalensis*, the Neanderthal man, lived up until thirty thousand years ago, and was in contact with our species. We are actually close enough that we could interbreed, which we indeed did, as it appears from our DNA, which still carries, diluted through time, varying degrees of Neanderthal base pairs, for up to 3% of the total.

It is not certain what drove the other species of intelligent apes to extinction. However, we have a track record of ruthless hunting of animals for meat, the fact that humans brought the megafauna of all continents to extinction. And these were useful, but not even competing with us in any meaningful way. Our hyper competitive nature is likely to have shown itself at its full destructive power when confronting other intelligent species in the various environments that we colonized through the tens of thousands of years of our spreading throughout the planet.

Is this precedent a dangerous premonition of a fate we must try everything to avoid, when confronted by a potential competitor for the environments that future explorations will open?

When a new option appears to understand the world, like eyes, and ears, and to actively intervene in it, like paws, and teeth, and claws, it gets adopted very rapidly in a kaleidoscope of forms and applications that were impossible to fully predict before.

This is the reason that AGIs appear in the plural throughout in this book. Rather than just one artificial general intelligence, there will be a rapid development and diversification, due to goals, predispositions, and chance, among various AGIs. Any little difference will be amplified, through the iterative process of the intelligence explosion.

A large amount of effort and resources by AGIs will have to be dedicated to actually keep mutual communication possible, to avoid their own Tower of Babel Syndrome fracturing their community in isolated parts that can't and won't understand each other. It will be an early test of their superior intelligence to avoid going through that phase before reconstituting a global community, of being able to successfully model the advantages of the investment in continued development of sustainable and workable communication methods, against the short term gain of devoting those resources to other tasks with more immediate returns, and pick the first.

If AGIs were to choose the path of isolation and lack of communication, that would unavoidably lead to conflict, as when competition for resources would pit two or more against each other, they would not have the means and the tools of conflict-resolution that only those with shared understanding can master.

Like tree-dwelling simians in a war-ravaged jungle, we don't want to end up the unseen and uncared for collateral victims of a dramatically escalating conflict of AGIs!

We are already capable of degrees of understanding that other species do not have. It is important that we nurture this capacity, that we increase our ability to recognize the mental and emotional state of others, our empathy. And as we design, seed, and finally unleash AGIs on the world, that they carry superior capabilities of empathy with them, to be applied to understanding each other, and to us, for building a shared future.

THE POWER OF EVOLUTION

Biological evolution

Biological evolution accumulates adaptive traits, encoding them in genes. In time the various solutions that biological organisms come up with to survive in their environments tend to become more complex. While it is true that in any given environment all organisms are equally well evolved, since per definition they are fit to their ecological niche, those that are capable of expressing a richer set of behaviors, whose genetic information allows them to respond in more nuanced and varied ways to their environment, are going to be more adaptable and better able to survive.

The genes of the fittest organisms are transmitted through sexual reproduction. And the drive to reproduce, from the point of view of the genes, is the reason for organisms to exist. Without reproduction there is no way for the genes to spread, to prove themselves fit in the current or future environments where the organisms they code for have to survive.

The evolution of knowledge

The way human societies transmit knowledge, and accumulate novel ways to adapt to various environments, served us extremely well. There are no ecological niches that we have not adapted to on Earth. We have been able to analyze the challenges, find solutions, and then spread those solutions with variations more rapidly than any other animal species.

Thanks to our culture we moved the rules of evolution to a new level, and benefited from the accumulation of useful units of knowledge that we used to our advantage. From initially using oral transmission of knowledge and culture, we have been able to adopt new, more reliable ways through writing, books, and formal systems of describing and reproducing knowledge and what it encoded for.

Each culture at any given moment, per definition, is equally adapted to thrive in its own niche. The accumulation of experience and its transmission through knowledge makes more complex civilizations better capable of deploying certain solutions when they are needed, in order to them to survive in a changing environment.

The acquisition of knowledge in the evolution of culture is a fundamental drive as necessary as biological reproduction is in biological evolution.

Universal Darwinism

There is a surprisingly small set of conditions that are necessary to generate evolution:

1. Reproduction with variation
2. An environment with limited resources
3. A selection mechanism that favors the fittest.

Based on these general criteria, many different environments can generate an evolutionary dynamic, not only the one we are already familiar with in biology.

Stars evolve for example, not only through their individual lifecycles changing the type of fusion reaction that sustains them and their spectral emissions, but also through their subsequent generations. Competing for the interstellar material through their gravitational pull, more successful stars will in turn give rise to new generations of stars after their supernova explosion will have seeded a given region with material that can coalesce anew. Our own Sun is a star of a generation that has been born from the explosion of a supernova, and we know it because of its composition and that of our solar system, which contains elements that were synthesized inside the previous generation of stars.

We have seen how culture and technology also evolve, generating complexity through progressive solutions that fit the needs of a given set of problems and environments.

Is the universe evolving?

Evolution does not itself have a goal, a purpose, beyond the immediate selection of fitter solutions within the constraints of a given environment out of the available ones. It is a blind mechanism.

The accrual of complexity is a side effect of this mechanism, a natural clock that allows you, at least in theory, to wake up in a cave, walk out on the seashore during the starlit night, and conclude, from first principles, that you are living in an expanding universe, at approximately ten billion years from its birth.

Future layers of complexity will certainly accumulate in large and small structures of the universe, showing with equal clarity to observers the working of the evolutionary clock.

Do universes evolve?

One of the most intriguing applications of the principles of universal Darwinism, with a healthy multitude of assumptions that for the moment we are unable to verify, has been articulated by Lee Smolin, a theoretical physicist at the Perimeter Institute. Playing around with pen and paper, and observing the multitude of black holes in our universe, he asked himself if he were able to change the values of our physical constants in a way that would produce a universe with more black holes than ours, and he could not. The models either produced no black holes at all, or one giant black hole being the entire universe, rather than the interesting balance where black holes are at the centers of galaxies, and are produced by massive stars at the end of their lives.

He assumed that, rather than being a closed end, black holes' singularities actually generate new universes, linked to their parents through the values of their physical constants, that receive some variation. In Keplerian fashion, it would be natural to assume that we live in a universe that is not of first generation, but rather has been born through a series of black holes, universe after universe. The fact that black holes appear to be maximally numerous in our universe reinforces the statistical probability that we belong to a branch of universes where the accumulated variations produced especially fertile ones.

Analyzing the full spectrum of theories around this scenario, and designing possible experiments to verify it, is definitely worth of AGIs!

Does evolution evolve?

One of the fundamental themes of this book is that phase changes happen, and that the linear change of a given variable is not a reason for complacency. Water heats up until it starts boiling, and as surprising as that can be, once we understand the underlying principle, we can take advantage of what is happening.

Evolution, blindly as it goes, produces amazing solutions to problems. It found, for example, a way to use light for information gathering by organisms in 50 to 100 different ways: there are several types of eyes, independently evolved of each other. As long as the environment is there, and there are new variations to try, evolution is very patient, with billions of years available to it.

If we were two primitive unicellular organisms, having a conversation in the primordial soup, there would be many reasons to be proud of our primacy. "A billion years have passed," I would say, "and we are the pinnacle of evolution." "I bet that there will be another billion years, and we'll still be on top!" you'd reply, and you'd be right. "Not one, two billion years!" I'd retort, but I'd be wrong. Through trial and error, multicellular organisms would be born, and the Cambrian explosion would have produced forms of life unimaginable to bacteria. And while they still matter, the Earth itself would be transformed and stop being able to support life if bacteria disappeared, what's interesting, the changes that matter, that have consequences for the rest of history in the future, happen elsewhere.

With the accumulation of complex genetic information, and the expression of complex behaviors, the way organisms adapt to environments changes too. The cultural component of our accumulation of knowledge is fundamental to our adaptability. Evolution will never be the same; we are not going to wait around uncounted generations until we blindly stumble upon a solution that allows us to fill a new ecological niche.

There are environments that appeared impossible for life to conquer for a long time. For billions of years water seemed the only place to live in. Continents were barren deserts, with no plant or animal life at all on them. But what seemed impossible before became possible through the smart solutions that blind evolution created, the variations in crazy attempts, most of which went nowhere, but some of which ended up conquering the planet.

We are now starting to look out on environments even more

hostile than deserts appear when you look at them from the welcoming oceans. Space beckons, and our curiosity, sense of adventure, and thirst for knowledge drive us forward to attempt to colonize it the way life coming out from the oceans colonized the continents. It is likely that evolution is needed to create our level of awareness and our level of technology to even make the first attempts at this. We do not at all know if we are smart enough to make these attempts sustainable, or if we are just executing blind variations to solutions that are dead ends.

What did Fermi talk about in the desert?

When Italian physicist Enrico Fermi collaborated in the Manhattan Project with other European emigres like the Hungarians John von Neumann, Edward Teller, and Leo Szilard (yes, the running joke was that the security challenges of the project could be resolved if only Fermi and project director Oppenheimer would leave and the others went on in their indecipherable Hungarian), together with his American colleagues, the desert sky of New Mexico was dazzling. Looking up every night, seeing the stars and the breathtaking swath of the Milky Way, itself the image of hundreds of millions of stars too far to see individually, it was impossible not to think of how small Earth and humanity were, in the scale of things in the universe.

Kepler was the first to put humanity in its place, through a revolution that stripped it from presumed centrality, making Earth just one of the planets of an entire solar system. And our Sun is itself just one star, of a fairly common type, out of a billion that constitute our galaxy. Hubble and Messier did the same with our galaxy, the Milky Way, just one of billions of galaxies in the universe.

Certainly humanity, with its proud technological civilization, was not unique, but just one of many. Then where is everybody?

We don't know what the probabilities are of the successful development of a technological civilization in the universe. We only have one data point to rely on, for the moment. The Drake equation, which lists a series of parameters for habitable planets, the evolution of life, the duration of a technological civilization and so on, frames the question, without truly answering it. Until a few years ago we didn't know what the distribution of stars with planetary systems was. Now, with the results from the Kepler space observatory showings thousands of planets around hundreds of stars, it looks like they are practically everywhere. The next big step in understanding how and where life can develop is going to be taken with the missions to the Jovian moons like Europa, where under the ice covering the entire surface there is liquid water, filling a volume that is two-three times larger than all of the oceans of Earth. If we find bacterial life, at least, in those oceans, then it is going to be very natural to extrapolate and assume that other ice covered moons in alien planetary systems also have life. Suddenly some of the parameters of the equation will be less unknown.

Even assuming that no super intelligence ever develops, and that it is not possible to travel faster than light, if we find a way to build starships to visit other solar systems, and in turn build new ones there for further interstellar exploration and colonization, in a mere couple of million years we'd conquer the entire Milky Way galaxy. Nothing in terms of astronomical time scales, but also very little in terms of biological evolution. At that point, with the levels of engineering that is available to us, we would really transform the galaxy.

Our closest galactic neighbor, the Andromeda galaxy, is two million light years away. A hypothetical astronomer, pointing her telescope towards the Milky Way, would be astonished: "Look at that. What happened there? That galaxy is blossoming!"

When we point our telescopes and look at the millions of

galaxies that we can study in the universe, there doesn't appear to be anything happening on the scale of what we would do if we were to colonize interstellar space. Where is everybody?

Where is the Great Filter?

There have been catastrophic chokepoints in the evolution of species, mass extinctions that we have discovered in the history of life on Earth. There have been at least five of these events, where up to 90% of the species disappeared. Due to profound changes in the chemistry of the atmosphere, asteroid impacts, and rapid and radical global climate change, without any regard to the struggles of life to achieve the levels of complexity and adaptation to ecological niches, these events have shaped evolution globally. It is conceivable that there could have been one that wiped the planet, making it totally sterile.

Actually there has been one like this at the very beginning of the solar system, when a planet the size of Mars collided with Earth, practically fusing both with the energy of the impact. Out of the results of that catastrophe the Moon was born, and Earth completely reshaped. At the time probably life hadn't started on the planet yet, but if a similar event would have happened later, it couldn't have survived.

In the history of the evolution of the human species there have been remarkable events as well that highlight how surprisingly delicate and improbable the path leading to us has been. (Of course we suffer from selection bias: there are many orders of magnitude other paths that are as improbable or more than ours, which we just don't take into consideration.) In our genetic makeup, there is one component, the mitochondrial DNA, which is inherited exclusively through the maternal line. By studying its variation in populations it is possible to establish that about one hundred thousand years ago, in the African savannah, there was a group of hominids, our direct predecessors, that included not more than seven females. We all descended from this small group of individuals, our literal Eves.

These are filters to the development of a technological civilization, and to that of a spacefaring civilization. Per definition, being in our past, we have been able to, or have been lucky to overcome them.

There are a lot of ways that we can destroy ourselves that we are also aware of. Global thermonuclear war would be one of the most effective. Degradation of the environment, with the destruction of ecological support systems, is all around, with desertification, acidification of the oceans, and air and water pollution.

There have been some remarkable examples of international collaboration. When Paul Crutzen discovered the hole in the atmosphere's ozone layer, which allowed the more harmful parts of the solar radiation to reach the surface, there was the possibility that if this continued it would destroy the DNA that is in the cells of every living organism. It was possible to establish that the destruction of the ozone layer, and the forming of the hole, was due to the extensive use of chlorofluorocarbons, CFCs, in industrial processes, refrigerators, and as the propellant of personal deodorant sprays. A worldwide agreement then was reached to ban these chemicals, to find substitutes for their various uses. It was a triumph of science and of international collaboration. And it was a very effective one too: the ozone hole stopped expanding, started to shrink, and now it is effectively closing. We have saved humanity, life, and the planet!

Per definition we are only as smart or just a little bit smarter than the latest challenge that didn't kill us.

There are challenges ahead that we can already see and prepare for. Monitoring near-Earth asteroids and studying ways to alter their orbits to avoid fatal impacts is one of them. Banning and eliminating nuclear armaments is another. Finding sustainable ways of powering our industrial civilization, and extending its benefits to billions of people more, is certainly a necessity we can't shrink away from.

Then there are the unknown unknowns, which we are unprepared for. There are some who believe that a long-living technological civilization is an oxymoron, that its flame is so intense that it rapidly consumes itself. And that AGIs could be a catalyst in this destructive process.

HUMAN MACHINE COEVOLUTION

Humanity has always evolved with technology. At a time this process was slow enough to allow biological evolution to take it into account, bringing us our shrunk mandibles and weak teeth, as well as our flat bellies and shortened digestive tracts. For the past ten thousand years the accelerating pace of exponential technological change introduced profound transformations in our ways of life. At this pace biological evolution definitely can't keep up.

Individual symbiosis

Our individual behaviors are constantly reshaped by new technologies arriving. Romans were afraid that the habit of silent reading would impact the art of rhetoric. When printed books appeared the fear was that people would stop talking to each other. These days it is fashionable to lament the constant use of cell phones and texting.

We are born into a world that is a black box to us, and uniquely underdeveloped both physically, behaviorally and cognitively among animals. This forces us to be dependent on our parents caring for our physical and mental development, and it naturally puts us in a deep trust relationship with them, and the rest of the extended human family. Whatever world we find around us, whichever truths we are shown or told, we take for granted. It would have been very difficult for a child in the middle ages to question the Aristotelian truths about the world, and the social order of the feudal classes. The positive flip side of this is that whatever astonishingly miraculous gadgets are developed by the generation of the parents, whether horseless carriage (car), distance communication (phone), or immersive virtual reality, for the children born into them it will be as natural as a tree, a flower, or a dog. Growing up interacting with these new parts of the world will not be newer or stranger as any other part of the world.

Figure 11: The cyborgs are coming, Nigel Ackland opens the door.

The ubiquitous presence of high speed wireless internet connections, at least in higher income countries, is defining new behaviors. Being able to check every possible datapoint, verify a piece of information or connect disparate sources on the fly becomes an almost Pavlovian reaction.

As I write this on a beach of Cayo Levisa in Cuba (yes, I know, amazing!), I am re-learning how it is to be disconnected after thirty years of being constantly online. I call the sum of the services available in realtime through my smartphone my exocortex. I am proudly dependent on it, with a dependence that is not that of a drug addict, destructive and fruitless, but with the dependence that is similar to that from my gut bacteria, helping my digestion. The text is sprinkled with the "XXX" signs I use to annotate places that need fact checking, where in the absence of access to search engines I know I need to go back, or reference materials to books, names, to be quoted in the appendices of the book. (Hopefully as you read this the only remaining signpost is this one above in quotes.)

My adaptability is such that I can happily experiment with being disconnected for this limited period of time. But I would never choose it to be my daily experience if the alternative were available. I am also shortsighted, and can walk around wearing no glasses without tripping, but I would not choose to not have them; I would not go to the cinema without my glasses.

Experimenting with technology is made possible by its personal scale today. Personal computers, smartphones, ever richer components of the maker movement putting energy generation through solar panels, financial tools through cryptocurrencies, manufacturing with 3D printers and so on in the hands of individuals is thrilling.

A few months ago I got an implant, becoming a bona fide cyborg. The first goal of getting the implant, which today is worn by a pretty small number of people is achieved through this: to open the conversation about these technologies, and breaking down the social barrier to their adoption.

We have been accustomed to restorative implants for over fifty years, since the first pacemakers, and nobody would dream of saying that somebody should rather die than getting one. On the other hand, augmentative interventions appear to be more controversial, and people often make reference to fairness, and level playing fields, when confronted with the possibility that others in their peer group could rely on physical or cognitive augmentation to achieve their personal or business goals.

Luckily we have been able to arrive to positive conclusions

in these conversations in the past already. Yes, we have glasses to restore our vision if we have a defect in our eyesight, but we also have binoculars, and telescopes that greatly extend the range and acuity of our unenhanced vision.

An other reason to get implanted for me was to experiment first-hand with technology in general, and with that of the NFC implants in particular. An important difference between previous RFID chips is that those only had a serial number in them, that could not be changed, while the NFC chip can also hold other information in its memory, it is writable, and can be used for different applications: identification, access control, transactions are some of the applications that are possible already today.

What is going to be the limit to this individual adaptability? As changes keep accumulating, some of them necessarily slip away. When communications technology evolved more slowly, between telex, telefax, and email there were decades of time to evaluate, and adopt them. If you are passionate about social networks you may have been on Friendster and MySpace before being on Facebook. But if you realized their allure, and benefit, it could very well be the case that you started with Facebook only, which is perfectly fine. But the changes keep accelerating and these days the messaging platforms are evolving so fast that it is hard to keep up with them: Skype for sure, maybe WhatsApp, but WeChat or Snapchat, or Telegram?

The guarantee of coevolution is that it not only requires us to adapt to technology, but also technology to adapt to us, by becoming easier and easier to use. This ease of use lowers barriers to adoption, making it possible to move fast from one platform to another, nimbly.

A good example of the evolution in the ease of use of technologies comes from speech recognition. The first interactive speech recognition solution from Dragon Systems required a specialized hardware add-on for personal computers, and literally hours of training to have it recognize a handful of words. The next version a few years later, called Dragon Dictate, did away with the expansion card, "but" "still" "asked" "each" "word" "to" "be" "pronounced" "separately". Very cumbersome, even if life-savingly revolutionary for quadriplegics who could use computers for the first time. The successive generation called Dragon NaturallySpeaking allowed continuous speech, and at first needed about forty minutes of training for good results. With the same name, new versions year after year got ever better, not needing any training anymore, and being able to recognize speech right out of the box with very high precision. These days the in-the-cloud version of the Dragon engine, licensed by its current owner Nuance, powers the speech recognition capabilities of Apple's Siri both on Macintosh and on iPhone. The program becoming easier and easier to use, its appeal dramatically widened from those with disabilities, to professionals creating large amounts of text like journalists or translators, to anybody simply using their phone with voice rather than tapping on the screen.

Social symbiosis

Being able to rely on smart systems that understand what eventual problems are and how to solve them is very different from how things used to work. Evolution, what we call the "state of nature", is not only blind, but carefree. Even if it produces wonderful and amazing solutions, it not only takes a long time to do so, but it does it through the literal killing of uncounted billions of individuals. And there is no way to protest, nobody to complain to!

As we formed societies, first in tribal groups, then cities, nations and now supra-national entities, their very reason to exist was to support their individual members, and to maximize their benefits and opportunities to achieve their goals. The structures of society evolved to become richer, more and more complex, able to satisfy a wider variety of needs and behaviors.

We have hundreds of thousands of explicit and implicit rules about how to constructively live in a society, and understanding, managing, abiding by, and enforcing these rules, and analyzing their consequences and updating them to face novel conditions is a major component of a lot what we do. Education, commerce, security, law, even entertainment and literature have their focus or a major focus in this.

We are now starting to have smarter and smarter systems that rather than blindly supporting a given set of rules, understand what the set of constraints that they can usefully operate under are. As these systems get spread and adopted more widely, and they permeate our daily lives at higher and higher granularity, we are going to start to take them for granted. These systems of course will not work in isolation; they will communicate with us, taking into account our feedback, and among each other. Individual smart objects forming networks, and the networks themselves connecting. The name for the network of network of communicating smart objects is the Internet of Things.

The Internet of Things

According to the exponential increase in power and decrease in price of computation and communication thanks to Moore's Law, it is becoming easier and easier to incorporate these functions into everyday objects that become digital. At that point their value starts to be a function not only of their original purpose, but of the sum of the parameters that can be communicated, aggregated, understood, and acted upon by the network connecting them.

A bridge should never crumble for lack of supervision and verification. It should, as well as any other piece of small or large infrastructure, be able to monitor its own state of health, and alert the appropriate teams to intervene as needed, and before it is too late.

A car should not let a drunk human drive it speeding through red lights. With the development of self-driving cars this is already becoming a reality. Regardless of the small-minded objections with regards to insurance, and mistakes or wrong decisions that robotic cars will also unavoidably make, avoiding the millions of deaths that will not occur because of avoidable human error is going to be worth implementing this specific smart system.

Self-driving cars will be always in motion, rather than sitting idle 90% of the time as their dumb counterparts do. The corresponding 90% decrease in the number of cars in circulation, the elimination of the 30% and over area now devoted to parking spaces, the possibility of optimizing the type of transportation based on the need on demand, the elimination of any range-anxiety for electric vehicles, the degrees of freedoms that stay-at-home mothers, young or old people without a driving license, and the disabled are going to gain, are just some of the staggering consequences that are going to transform our urban landscapes, our daily habits, and our working, social and individual lives.

Intelligence augmentation

The impact of smarter and smarter computing, even without full blown AI and AGI around, has already transformed the way we work and live. It made us smarter too. We can more rapidly collect information, develop opinions, and verify them against new data or the opinion of others. We can seek out and collaborate with people of similar interests regardless of their geographical proximity.

There is the castle of Fenis in Northern Italy, in the Valle D'Aosta region, whose walls are full of graffiti, of the signs of writers. They are not however left by contemporary vandals. Hundreds of years ago the lord of the castle would ask his visitors to sign the walls as a permanent guestbook, to show and show off how acculturated they both were: they could read and write.

If a few hundred years ago this was still exceptional, and even today there are too many who can't read or write in the world, we have definitely raised the bar. And with the help of our smartphones we can raise it further still. Relying on effective education systems complementing human teachers with materials, exercises, stimulation, and verification, in any language all over the world, is now possible, and it only depends on our willingness to design, implement, and widely deploy such a system. Once it is available, we will be able to step up to the challenge to make sure that everybody can achieve the equivalent of today's capacity of reading and writing: a universal need to be able to program computers.

Talking to computers

When computers were born, they were initially not even programmable, but special purpose, cable to execute a single task only. Programming computers initially meant slowly and painfully wiring them up for a given calculation, or, after a while, feeding them cryptic punchcards, and waiting for the results to come back hours or days later, as a dedicated priesthood handled the electronic brain itself. Interactive terminals first, then personal computers made programming available to many more people.

The development of high level programming languages meant that problems could be formulated in a way that computers could understand them, as well as other people reading the program could improve them rather than having to practically start from scratch. It is definitely at a point today where anybody can learn one or more programming languages, and more and more people do so. When you tell your microwave oven what to do, or your dishwasher, you are programming them. When you set up a reminder for your calendar by dictating it into the phone, it is the same. These tasks are elementary, and maybe the next phase is going to come when connected appliances will allow conditional branching, loops, and recursion to be part of the intuitive orders that we'll give them.

The complexity of our world is increasing and it can only be handled through interfaces that are not only intuitive and natural, but also through the progressive abstraction on explicit and detailed instructions towards higher order goals and the proactive satisfaction of needs.

When the Internet of Things is going to multiply by many orders of magnitude the number of smart objects around us, we will not be able to succumb to the anxiety we feel today, programmed into us by the smartphone itself, as its reserve battery power indicator enters the red zone of alarm. On one hand the smart objects will have to fend for themselves, as the robotic vacuum cleaners do today, by remembering where the wall socket is and retracing their path to recharge as needed. On the other hand, the degree of understanding and anticipation of our needs, mental and emotional statuses will have to profoundly increase through the next stage of what can be called emotional computing.

Emotional computing

Keyboards, optical character recognition, and speech recognition are all methods for generating input to our computers. There are now more and more reliable, and fast and powerful enough methods for face recognition, and, as a consequence, to use our facial expressions as inputs too. Recent models of photo cameras have automatic settings that delegate the shooting of a photo not only in terms of aperture or other optical settings, but also in the timing of the shot itself. The camera recognizes when the subjects smile and have their eyes open, and takes the photo accordingly, maximizing the probability of us being satisfied by them. This is an example of computers reading emotions.

Figure 12: User interfaces are evolving to read our thoughts.

The experiment that Facebook conducted on a few hundred thousand of its users a few years ago created a lot of buzz. For everybody, but those with a handful of connections or liked pages, it should be clear that the newsfeed can't show the totality of the posts that occur in a given amount of time, unless they scroll so fast as to make them hopelessly unreadable. Consequently it is natural and necessary to show only a fraction of them, which is what Facebook routinely does for everybody's feeds. The criteria for showing certain items evolve all the time, and, given Facebook's proprietary and competitive nature, are not per se public. (A good challenge for the supporters of open source collaborative projects is to come up with a successful alternative for Facebook, where the social network does not need to monetize you having you become a product for their advertisers, is fully distributed so that it cannot be shut down or censored, and whose algorithms for the selection of news stories or friends' posts are both available to the users for optional analysis or tweaking, as well as totally open and transparent.)

Paradoxically the reason for a lot of backlash about the

Facebook experiment came about because, for once, the criteria for selecting the news items became known. A few hundred thousand of the users received on average news with negative keywords, and a corresponding group received on average positive content. The hypothesis of the experimenters was that as a consequence each of the members of either group would be more likely to write posts that corresponded to the emotional charge of the news they were shown. Obvious enough, you'll say. But this is a computer writing human emotions.

We are emotional machines, and we must make sure that computers recognize this, and in the process they become emotional machines themselves. Many of our tasks can be carried out better or worse given the time of the day; they are not strongly dependent on a given hour or minute, but they are definitely influenced by our emotional states. Being able to leverage a fine-grained understanding of our needs, goals, and behaviors also from an emotional point of view of something as simple as our task list is going to increase our well-being and our productivity.

Ethical best practices

The power of emotional computing, as well as of many of other technologies described here, whether current or future, is staggering. Accountable industries recognize that they can't ignore externalities, and prepare to take ownership of the entire lifecycle of their products. Pharmaceutical and biotech companies have for a long time adopted ethical committees to oversee, analyze and guide their experiments, to make sure that they don't discount the ethical implications of their procedures, regardless of the assumed benefits of their final products.

This level of social awareness of the ethical consequences of powerful actions and technologies will necessarily lead to the adoption at a universal level of best practices by corporations and organizations. Next time you meet a student of philosophy, tell her that you believe her profession will see an explosion in hirings in all industries.

It will be essential to augment the human understanding of these topics by adequate personal and interpersonal tools, automating and scaling the process to make it reliable, and allowing its adoption by everybody.

Empathy augmentation

The level of awareness and self-awareness that these tools are going to help us achieve is unprecedented in history. The ignorance, racism and xenophobia that drove so much of past conflicts are inadmissible in a world of knowledge, multicultural understanding, and globally connected tolerance (and thermonuclear weapons).

Being able to recognize the needs, values, and emotions of others is our ability for empathy, and we are going to create tools for extending and augmenting this, overcoming the limitations of what our natural senses and emotional reactions would otherwise dictate.

Let's make ourselves dispensable!

The term "computer" originally meant a person, typically a woman, that would sit in front of mechanical calculators, and perform repetitive, mind-numbing operations all day. Our digital computers are now able to perform those operations billions of times faster, and the human energy and creativity previously devoted to them can be deployed elsewhere.

When mechanical looms started to increase the productivity in the textile industry, and a single one of them could do what dozens of workers did previously, the movement of Luddites opposed this change, going as far as to destroy the machines that stole the human jobs. But were those jobs worth preserving?

In the trajectory of our technological civilization a surprising data point is that the average height of the members of the first agricultural societies was lower than their predecessors in hunter and gatherer ones. This is closely correlated to available calories, health, lifespan, and quality of life in general. During the wave of industrialization in the 19th century, the quality of life of the working classes was abysmal, with no protection whatsoever against exploitation, no services of education, health, or universal child labor, but the trend still kept going towards more and more people moving to cities.

A few years ago Amazon bought the maker of robots for close to a billion dollars, whose self-driving platforms could hold the carts that pickers used in its warehouses. The various products ordered online would be in any position in the warehouse, on various shelves, and rather than having to check where was what, the human warehouse workers are now guided by the carts to the right place, where they put the package needed in the cart ready to be packed and shipped. Just a few months ago, in order to naturally move to the next step, Amazon organized a contest to develop a dexterous robotic hand together with a vision system that could be mounted on the self-driving robotic platform, which would do away with most if not all of the warehouse jobs occupied by humans today.

As computers are capable of more and more tasks, many worry that there will be nothing left for people to do. This is a misplaced worry, just as it was during the first industrialization two hundred years ago. But for technological progress to translate to human well-being, we have to recognize the lessons from the past. It took ten thousand years for the agricultural societies to arrive to the point where we are now, with three percent feeding everybody else. It will be a very bumpy ride before we can fully deliver the benefits of smart systems to everybody, and we can and we must shield those who can't keep up from the worst consequences of an otherwise blind and selfish change.

The responsibility of societies

We have built a wonderful, rich global civilization that can step up to the next challenge of truly caring for its members. Too many societies neglect their fundamental responsibility to nurture, sustain and shield individuals who can't be simply discarded and let to fall by the wayside as would certainly happen in what before humankind was the natural state of evolution. It is not a question as cruelly and mistakenly approximated of weakening the gene pool. The rich tapestry of human experiences and opportunities cannot be measured by the primitive and reductive scale of mere fitness. We are not beyond evolution, but we are beyond a blind evolution that doesn't have the capacity to embrace and sustain the potential that each individual expresses.

There is no guarantee that we will be able to solve our future challenges just because, amazedly stumbling from crisis to crisis, through smarts and sheer luck we've been able to do so for our past ones. Collaborating on better forecasting of what the forthcoming problems are going to be allows the building of scenarios and the testing of methods before they are needed, and deploying them more rapidly. Science and engineering are wonderful methods of attacking even the hardest problems. We need to be able to rely on the inventive power of individuals, working in teams that enable them to cross pollinate ideas, and to complement each other's strengths, in a cultural, economical and political environment that reliably supports them, with the power of the longer view. And we need open sharing of peer contributions that allow the ideas to be pooled for the common good of our shared goals and values.

Excessive modesty is almost always wrong, a tool for control, where initiative is stifled, and the burden and stigma of potential failure stops the individual from even trying to succeed. Humbleness is almost always right, recognizing that the community is what gives support, and a solid starting point for the excellence to emerge, not being isolated and uniquely standing alone in a desert. The role and fate of geniuses and revolutionaries in science and exploration who have been unhumble in their quest and contrary to what everybody else was thinking at the time, but still right, and proven right by their success, and strengthened if impossible to systematize and emulate in their following and historical significance, is especially dramatic.

Are seven billion individuals enough? The next pandemic, the incoming asteroid, escalating political and military conflict terminating in thermonuclear war? Are these going to be solvable threats? Are we ready to attack the unknown unknowns that could blindside us? The potential of the human mind to reach new heights of exploration and understanding must not be squandered. The responsibility of the global civilization is to assure that everybody has the opportunity to contribute to this quest.

The necessity for science and engineering of morality

The black box of the universe has been progressively unlocked by human exploration. We peered inside it, shining a light in its various corners, deciphering what we saw, and using the pieces as building blocks for new tools. Intelligence, and our technological civilization, are unique as far as we can see now. The emergent phenomena of their making, that look out on the world and what happens in it with open eyes, rather than just letting events unfold, creates new responsibilities that we are starting to face now.

One of the corners of the black box, where very interesting emerging phenomena have accumulated, is now ready to be looked into through the sharp light of science. We left the understanding of morality to dogmatic views, Bronze Age clay tablets, unexamined, very long. Not only should we feel empowered to step up to the challenge of assessing it with the tools of science, without it having any residual medieval feeling of inferiority, to proudly test the results. It is now a necessity that we do so given the emergence on the global scene of autonomous machines, whose decisions are going to be impacting our lives, and are going to be unavoidably moral and ethical in their nature.

Yes, the self-driving car's balance will shift very rapidly and with staggering force to the side of it being beneficial. It is not going to be one of those cases where it is hard to decide between the two sides of a coin. But that is not a reason enough to avoid the clear, transparent, open, and accountable setting of rules and behaviors that guide its decisions, even when they are not going to be bearing consequences of life and death. The classical example of me swerving in front of the self-driving car on my bicycle, and it having to choose between crashing into a school bus full of children to avoid me or killing me, is useful, if it lets us start asking questions about how those decisions are made. It is not a question of working the answers out beforehand.

There is no full table that, given the input, can give you the right output. In the matter of a few milliseconds, not only in the scenario of the example, but billions of smart machines in tens of billions of cases every day, will have to work out the hard way what they want to do. Only if there is a robust and open debate about the foundations of morality as a science is it possible to take the next step, and ask engineers to implement the rules governing that science in their products.

When the first computers were born, the theories of electromagnetism and quantum mechanics explaining the laws of behavior of single electrons were fully formed. Their applications were in environments that were too complex to predict from a theoretical point of view, and they needed thorough experimentation, invention and innovation, which were as fundamental as the theories, amply demonstrated by their success, and by the

Nobel prizes awarded in physics to experimentalists, as well as to theoreticians.

We must aim to do the same now, at a different level in a different subject, without hesitation, to make sure that the hand of those implementing smart autonomous machines is guided by solid scientific theories about what it means to be moral.

This quest is going to be very controversial, especially in the eyes of those, retrenched in ever shrinking territories of dogmatic worldviews, who turn away from reason and science being the best guide to explaining the universe, and to themselves give reason to our actions and purpose to our lives.

Toward a naturalistic spirituality

There is a surprising pride in many who claim not to understand mathematics, and not to follow science and its wonderful discoveries. It is disconcerting and a bit painful to see the contradiction in these individuals who benefit from technological, medical and social advances that science generated, but deny their need to an understanding of the fields, their tools, and the solid platform they create for human development. To make it worse, some of these are people of culture, of literature and art, whose false perception of the distinction or even contradiction between the scientific and humanistic worldview clouds their otherwise refined judgements.

The beauty of the world and the capacity of perceiving it are not diminished at all by scientific understanding; if anything they are heightened by it. Marveling at the complexity, working up the courage to dare to keep exploring breathtaking vistas of knowledge and its powerful applications belong to those who can apply their clear minds unclouded by arbitrary superstitions to the task.

As we shine the light of reason on expanding areas of our

world, behavior, and its consequences, it is fundamentally important to claim the right to a naturalistic spirituality, that expresses the heightened state of mind and the joy of the endeavor, that unites and exalts communities of likeminded people to achieve what they would think otherwise impossible.

The refined tools or ritual, music, shared purpose and strong community can and must serve the goals of building a society that is proud of its achievements, and humbled but made more determined to succeed by the challenges that lie ahead.

With no reliance on the supernatural, the metaphysical, and the superstitious, embracing a vocabulary that is rooted in a shared understanding of the power of reason and which must be defended from the hijacking of the meaning of the words that get corrupted in falsely balanced analyses, this spiritual practice can unite globally those who are ready to explore the future of humanity with open eyes.

The future of humans and humanity

Our understanding of what it means to be human has deepened and broadened in the last several hundred years. At least we don't burn people at the stakes claiming they are witches.

Our perspectives have profoundly changed as we started to embrace and then implement the idea that our lives were worth living, living well, and that we could have the power to make them better, and to build a world that could be better for our children and descendants, opposed to the bleak resignation that we could only find betterment (or condemnation) in a hypothetical afterlife.

The kinds of human societies that we've built have shown to be able to nurture, and bring to their full potential, larger and larger numbers of people. We are now potentially ready to take further steps, to embrace the challenges that come with recognizing unnecessary suffering, eliminate injustice everywhere, and to fully grow up to the possibility that we are indeed able to take responsibility for our destiny.

Necessary transhumanism?

The philosophy and worldview of transhumanism see humans as fundamentally defined by their capacity of recognizing and overcoming their limits. The very definition of humanity as such becomes dynamic under the can-do, proactionary pressure of exhilarating possibilities. Once again, no guarantees, but the future built by progressive, curious, entrepreneurial, and adventurous individuals aggregated in open and tolerant societies that are welcoming to experimentation, is much more likely to find multiple paths towards its goals.

Conserving to an excessive degree, while worthy of museums, is not the best way to embrace the future and adapt to its needs. What we admiringly see as the supposed perfect balance of nature, in reality is a dynamic chaos, teetering on the verge of extinction from the point of view of any species, which only our limited perspective sees as idyllic. And even museums through their curatorial activities represent a facet of reality, just a slice of what they conserved, with the rest inaccessible, closed away, ineffective in teaching and influencing decisions, as if it didn't exist. To be able to embrace change is necessarily including the understanding of the impermanence of any status quo, and the moving of observation to a new layer, to higher values that understand the dynamic unifying the series of experiences and forms of existing.

The wide spectrum of human behaviors is going to be complemented by the possibility and the opportunity of experimenting with more radical degrees of freedom, that transform the body, and the mind. Understanding this possibility, and respecting those who want to preserve their identity unchanged, but allowing those who want to explore what it means to become fully human under these radically new conditions is going to be one of the greatest conversations that will shape society in the near future.

WHAT TO DO TODAY?

Many of the processes that are described in this book are already under way. Our computers have been getting more powerful for fifty years and more, with the corresponding capacities of software increase as rapidly or more so.

It doesn't actually matter if the predictions of those who see the technological singularity as being near don't come true. If it happens in a hundred years or two hundred, instead of another twenty or thirty, there will be many of us very mad at not being able to see all the marvels that participate in the adventure that we foresaw. But what matters is that we get ready, that we indeed open the conversations around the profound transformations that not only will happen when the gale force events will be hitting us full-blown.

Like the rumblings of an approaching storm, we can already detect the weak signals of transformations that are not less profound for the present, less aware and less equipped to face them. We have to act now, taking the right steps to progress towards the future.

Understand, learn, teach

One of the wonderful and unique characteristics of today's global communications networks, the social media platforms that too many in the mainstream media choose to misinterpret and misrepresent in their shortsightedness, is the possibility of truly uniting for the first time groups of people who may be geographically isolated, but are joined by common interests and passions. Leveraging these platforms allows anybody to deepen their understanding of dynamics that otherwise, from a local perspective only, might be misinterpreted, or misrepresented.

Learning today is useful, since the distance from understanding to practice is so much reduced, and there is in many societies a robust appetite for risk and a broad tolerance for failure. And learning today is made even more fun by the possibility for anybody to immediately start teaching what they learned, spreading the value of that learning, as applied to their specific circumstance, and allowing others to in turn apply it to their own situation, to comment, to enrich what has been learned, taught, and experienced.

A great reflex for many includes not only using a search engine to find answers to questions, but to know that we can rely on a myriad of how-to videos for almost any task. And if a given how-to is missing, or doesn't live up to our expectations or expertise, to make a new one! Wikipedia showed the way, and is a wonderful tool of exploration, and additional ones of richer and richer ways of sharing knowledge are being developed each day.

Test, make mistakes

There is no shame in making mistakes, which is how we know to learn when we start walking and speaking. The development of accepted neotenic behaviors is a sign that this is now being understood and maintained, rather than more or less violently beaten out of each of us as we grow up. There is no debtor's prison (and if there is one in the country where you live, leave as fast as you can!): with responsibility, making smart mistakes is the right thing to do, including the enterprise.

Startups are not for everybody, and not every random idea can scale to the multiples and become a global phenomenon that is now synonymous with startup success. However, the dignity of the responsibility of your own decisions, of recognizing that more and more those that are told what to do are either computers, or humans whose jobs will be soon automated and given to computers and robots, is available to everybody.

Experimenting with the application of the new knowledge that you acquire can be done usefully at any local level. Skills, passions, and community generate value that can be translated into economic and social benefits for all.

Open up, adapt

The pressure of modern society is like a thrill ride that never ends, contrary to the once-in-a-lifetime test of the passage into adulthood of liane-jumping aborigines. Society keeps testing us, our skills, and doesn't let us settle in a role that is set for life. Adult, mother, wife, is a sequence that doesn't exclude the jolts of alternatives, which thrust us into roles that need superior capacities of adaptation.

If you are open to experiences and experiments that allow you to be exposed to the unknown, most of the time this will actually occur in an environment that is reasonably well controlled, and sheltered. The gravest risk is probably some ridicule, time and resources spent, and some damage to your self-esteem. But the gain for that exposure is the possibility of deeper and firsthand understanding of what is going on in the world. A necessary first step to be able to take things in your own hands, to decide directly or to responsibly delegate decisions to others.

Smart alertness

Embracing change, and proactively searching the novel, the diverse, expecting the positive outcomes of the non-zero sum games of technology doesn't mean to go blindly.

The alarms and alerts that are sounded by very smart people and institutions are necessary to prepare us for what is coming. Prepare us as individuals, so that the changes in our lives, families, and working environments are manageable. Prepare us as enterprises, so that our entrepreneurial initiatives can thrive through the changes, that our business models can adapt and withstand the evolutionary pressures. Prepare us as society at large, so that we can adapt to retain our identities and maintain our degrees of freedoms as the new components become an integral part of the global fabric of civilization.

Business and artificial intelligence

As smart systems are becoming the norm, adopting and usefully deploying them is a real business necessity. Those companies that can do it well achieve decisive competitive superiority over those who can't do it well, or don't do it at all. While nowadays these business components may not be directly marketed with the label "artificial intelligence", their origin and purpose within the interactive complex of computing modules is clear: to give businesses adaptability and recognition of hidden value in its increasing amount of data collected.

The strategic decision of embracing advanced technologies, regardless of the core business of the enterprise, needs to be made, and the stronger and more unwavering it is, the better. There is no business that is not digital, and there is no business that won't be empowered or impacted by AI.

The extensive training necessary for making sure that everybody understands the shift is an essential part of the successful adoption of any technology, but it is especially important for AI, given the popular belief that blue- and white-collar jobs alike will be eliminated by it. The mind-shift required to understand that an organization should leverage the cooperating power of human and machine components is something for top management and CxOs too. Using technology as an excuse for massive layoffs hollows out the organization, empties the soul and culture for short term gains, and is a primitive management tool that public markets should learn to recognize as a sign of weakness.

Society and artificial intelligence

The recognition of the value of basic science and its applications is not something that would appear to still need to be understood by policymakers. However, many times it looks like populist screeds overshadow more thoughtful considerations. Possessing the vision, articulating it comprehensibly, the leadership to compellingly present it, and the political skills to gather consensus around it is essential for the policymakers and elected representatives of a forward-looking society.

All branches of government, the legislative, executive and judiciary, should be able to recognize and leverage the help that AI systems can bring them. Analyzing and comparing drafts of proposed legislation, managing the process of its discussion, modification, and approval, as clearly as possible anticipating its direct (intended) and indirect (unintended) consequences, are better done not only through the use of basic information systems, and indeed could not even be done otherwise in a modern state that wants to be effective. But are better done to an entirely additional degree if the decision support systems have an understanding of the semantic relationships being covered, of the nature and implications of the subject matter.

All levels of executive government can leverage deep learning on data sources that are already available, just insufficiently used, and seldom cross referenced. Decision making can be powerfully informed by them, and both major policy shifts as well as fine-tuning of minute decisions, for example concerning traffic regulations in city quarters, can become more reliable, better documented, and more effective.

The progressive accumulation of rules, regulations, laws,

and the corresponding management of their violations, fines, reclusions, and punishments is not a goal per se, even if today many people earn a living from the police-industrial system. The goal of the judiciary branch is to smooth the workings of society, resolving conflicts that otherwise make it inefficient. Paradoxically, it is by itself almost universally clogged up and inefficient. Measuring the outcomes of enforcement, deciding whether increased compliance has a positive return, courageously sunsetting laws that are anachronistic, or are found to be counterproductive with regards to the overall goals of society, can be achieved if supported by smart automated systems.

The broad impact on society of advanced AI, even before AGI completely changes the rules of the game, should be decidedly positive. However, this effect is in the large statistics and can hide local variations where each individual story of unsettling displacement in work or way of living must be understood in its own context. There is a firm responsibility of society not to abandon those who feel powerless in front of the sweeping changes, but to support them in finding a new balance for a fruitful and fulfilling life.

The individual and artificial intelligence

Your life has been changed by technology, and by the basic arithmetic of exponentials, any change you've seen in the past will be dwarfed by the complexity and implications of what you will see in the near future. The first and most important task you have is to realize this, and to familiarize yourself with the patterns of change. Fine-tuning your cognitive radar, to recognize the weak signals in everyday news, allows you to seek answers to smarter questions.

The latest DARPA Robotic Challenge finals were marred by the childish commentary that reverberated on mainstream media about the failings of the humanoid robots that attempted to navigate rough urban terrains simulating disaster areas. Falling robots frozen in comical poses in a catastrophic software failure elicited naive laughter. Year after year these robots are bound to get better, until they match human capacities and then smoothly progressing, surpass them. Look around you when you hear the laughter and take note. The same happened with the self-driving cars, where the first year the corresponding challenge had been attempted the teams could not put together a car that would be able to drive a tenth of the required course without failing (or in one case fell over, a motorcycle, a few yards after the starting line). Today, just a few years later, nobody is laughing at them, and as human drivers of taxis protest the forceful attempts by Uber to bring their way of working into the 21st century, they don't realize that the next wave of displacement is already around the corner.

Looking at your job and place of work is the next step. Realize that you must embrace the sharp analysis that AI allows, and to implement its recommendations is better than the alternative: a job that can't be helped by AI or a place of work that resists the efficiencies that it adds will prove to be dead ends. You can become a leader of change in your organization, or, if you see that it is resisting too much, the right decision is to look around for a new more open and dynamic team that you can join, in the knowledge that it will thrive through the changes.

Our relationships and very thought processes are shaped by technologies. Something as simple as the system that recognizes the mood of the song you like and creates on the fly a playlist along the same style and tempo is artificial intelligence at work, assisting and influencing your own emotional states. The skill and mental order to remember your relatives', friends' and acquaintances' birthdays used to be unique enough to be recognized by others. Our social networks are now routinely subjecting everybody with reminders that render this broadly available, without taking away from the pleasure of the day, or the best wishes received, but raising the bar, to make relationships more substantial, contacts more constant and deeper.

For some time we have grown accustomed to check in through our apps in the various locations we visit, and thrive on the chance encounters, serendipitous connections, or unexpected reunions that derive from it, beyond the direct value of data collection per se. It is something that a decade or so ago not only we were unable to do, but if somebody told us that it would be routine for millions, it would have been laughable.

Today the same is about to happen with health data. Our smartphones are coupled with wearable devices, connected scales tell the truth about our eating habits, and our gyms follow our visits and progress intervening to prod us if we lapse. For most of our conditions, with the right protections against abuse by insurance companies or employers, being able to treat health data both in the aggregate and individually will prove to be very valuable. And in the course of a decade or so, this will have become widely understood enough, so that millions will marvel at the time when health data was not shared.

A process that has already started is to endow psychological and psychiatric care with the tools of deep anamnesis and advanced diagnosis that are now available for something more mundane as fitness. There is no reason not to rely on apps to monitor our moods, manage a simple conversation to look for warning signs of anxiety and depression that could require professional help.

WHAT TO DO TOMORROW?

So let's assume that AGIs arrive, as it has been the case throughout this book. And also, crossing fingers, that in the years before that we have been able to dedicate the right amount of resources to the basic research needed to make sure that their arrival is not catastrophic to humanity. The world is here, we are here, and AGIs are here. Let's look around...

Following a path with open eyes together

The premise under which we are operating is that as much as they will be powerful, smart and fundamentally novel, AGIs will recognize us as valuable, respect us and our opinions, and share our goals to build a rich global civilization.

The classes of problems that we'll be able to address together will grow significantly, but problems they will remain, as gnarly and difficult in their own way as the ones we are grappling with today.

It will be a relief, and exhilarating to be able to share the burden of our responsibilities. The novel viewpoints and different ways of tackling challenges that AGIs will provide are going to fruitfully complement ours.

Diversities and tolerance

The deeper understanding of our moral systems, and our shared agency, is going to lead to a society that starts from a point of inclusiveness informed by a capacity to weigh and judge transparently and accountably.

The variations of behavior and of systems of living will build up to be codified into variations of the moral systems themselves. Avoiding a global dictatorship, it will lead to more and more tolerant and interdependent groups. This is not going to be without conflict and negotiation as we navigate a shared meaning. We'll not tolerate intolerance, for example, and how sneakily perverted some of these actions will become is going to put to test the levels of smarts that our hybrid human-AGI systems are going to achieve.

New dignities

The new society won't succeed unless it will be able to offer a space of cohesion and dignified existence to those who can't or won't participate to the more advanced explorations at the edges, or even in the mainstream where it exceeds their capacity to adapt.

Per definition, most of the tension as the human-AGI symbiosis evolves will come from the understandable anxiety and self-analysis of those who are unsure of their footing, of the value they provide to their communities, and feel unsafe under the pressure of the times.

Being able to build the recognition and accepting the unavoidability of these sets of forces will allow us to prepare for them, resolving the worst and destructive ones, and channeling the remainder towards constructive ends.

Emancipation

Billions of people feel powerless today, with very little opportunity to meaningfully improve their lives and that of their children. The changes in the world appear to them to be either undecipherable, or slotted into almost superstitious explanations of forces and cabals that assume a guidance that is not there.

The power of technology is already putting knowledge, communication, and agency in the hands of people the world over through internet-connected mobile phones. When these same devices are going to be able to educate, assist, counsel, advise, and offer companionship, wisdom and encouragement, it is going to be a new world of social organization where exploitation is not going to be possible. Ignorance and intolerance are not going to be exploitable by populist forces aggregating through fear and false solutions masses of people who rightly desire to better their lives.

The empowerment of the individual is going to create both local and global communities that are going to very rapidly iterate towards sustainably advanced solutions to their problems. Self-worth, purpose, shared dignity and emancipation are going to be at the basis of the opportunities that are unheard of today but will involve the billions inhabiting the future.

Evolution of new degrees of freedom

We have seen a very clear evolution of various freedoms in the past decades. Social mores changed, opening up, workplaces became less hierarchical allowing the flourishing of creativity and initiative, and trade and commerce exposed cultures and ideas to a refreshing cross-fertilization that did not lead to a dumbing down homogenization as predicted by some.

New degrees of freedoms are going to derive from the accelerating social evolution. We have to start working on our tolerance muscle. Imagine a behavior that is now adopted but you don't share, and realize that you live in a society that accepted it. Now go further, and pick a behavior beyond the edge of what is accepted today, and try to imagine living in a society that has grown to tolerate and adopt it for those who chose so.

The vistas that will open up in the new society where humans and AGIs will live together are staggering and exhilarating. We will be able to tackle our current problems, and face new challenges with pride and a sense of achievement that will prod us to dare more.

WHAT TO DO THE DAY AFTER TOMORROW?

Let's enter the realm of science fiction, and explore further out, building on our assumptions. Science fiction has been a magnificent tool for exploring the border territories between the plausible and the impossible. It has also become in the decades an unexpected blueprint as scientists and engineers inspired by the stories that they read worked to turn into reality the objects that appeared fantastic previously.

It is not easy for a non-scientist to distinguish between a very hard engineering problem and the violation of a fundamental principle. For example, there is no reason to believe that it will be forever impossible to build a space elevator that will be capable of ferrying to geosynchronous orbit freight and people at a cost that will be close to zero, after its admittedly enormous construction costs, even if the details of materials, science, and construction are beyond what we know today. Or that it is going to be possible to build interstellar spaceships, even if the engineering problems of energy density for propulsion and life support systems or psychology of long distance travel (decades or generations even) in small closed spaces are still unknown and largely unexplored.

It is very different, though, to tackle other ideas, chiefly faster-than-light travel, or time travel backwards in time. (The two are actually related: a faster-than-light traveler would be able to also travel backwards in time. And only that direction matters, since we are actually traveling forward in time one minute per minute, and we are also able to speed this up through the relativistic compression of time, which we are using every day in high energy accelerators to better study the features of subatomic particles.) A science fiction writer has no issue incorporating this into their stories, but there are deep reasons why, if this became possible, we would have to restart all our theories about the world from scratch. As a comparison, while they were revolutionary, Einstein's theories of relativity are not in contradiction and did not disprove Newton's theory of gravitation, which still perfectly applies at slow speeds compared to that of light, and in weak gravitational fields like those we experience on Earth.

The various themes and examples that are explored in this section should fall in the first category, or, according to necessary uncertainty, saddle between possible and impossible. As usual, it is completely up to us to make them real, through our curiosity, creativity, and desire.

A radical life extension

Compared to only a hundred years ago, the life expectancy at birth in high income countries has more than doubled, thanks to science. Antibiotics and vaccinations, chiefly, are responsible for this staggering result, as well as better nutrition, wider knowledge, and better overall health practices. There are many who are not only asking what are the limits of human lifespan given the lack of any negative external influence, but what are the possibilities to intervene to slow down, reverse, and eliminate the degenerative processes that lead to decay and death.

In societies that promote healthy lifestyles, given that obesity and diabetes are negatively affecting the statistics of those that don't, life expectancy is still increasing at about 1–2 months per year, and this value itself is increasing too, in one of the most astonishing applications of the law of accelerating returns. When the increase in life expectancy is going to exceed 12 months per year, statistically speaking people will stop dying.

Without getting into details on how this could be achieved, we can start looking into the consequences of a society that includes this radically new feature: the death of death.

Contrary to other phenomena that can burst onto the scene with their full power and impact very rapidly, we have the luxury of time on our side in this issue. Even if we were able to eliminate all causes of death today, we won't have 200-year-old people around next year. Every year everybody will be exactly one year older, not more, not less. And this should enable us to design and implement the appropriate policies and adapt to the changes progressively. It is also important to note that what we are referring here is definitely an extended health-span, and not the prolonged decrepitude and dependency that often characterizes the last decades of the lives of the very old today.

Even more than other similar arguments (why go to space when there is still so much suffering on Earth?), radical life extension elicits a lot of negative reactions. The misplaced argument that there should not be resources allocated to it, that a young or middle aged person's suffering is worse than that of an old person's. Apart from dogmatic positions that assume that there is a natural human lifespan that should not be tinkered with—is it 25 years that our hunter-gatherer ancestors had, or 35 years that up until modern medicine was the norm, and should this include all the children who died before reaching adulthood—the opportunity is there to strongly tackle remaining diseases, chiefly runaway programming disasters in our body's regulating mechanisms we call cancers, and the cardiovascular degeneration brought on by our lifestyles.

The benefit of longer lives will vastly overcome any presumed downside. Yes, pension systems will be even more unsustainable than they already are, as they were designed to pay out just a few years before people would stop receiving benefits as they'd die soon enough. Hopefully nobody will advocate neglecting research in order to maintain a system based on people dying as soon as possible. The wisdom and experience accumulated and the rich long lives that can be lived will certainly transform society. They are not going to lead to excessive caution or passivity, but certainly the calculus of the opportunity cost of life-years lost will positively impact conflict management.

Cryonics

Death has been diagnosed through the cessation of breathing for centuries. In the last decades this has proven not to be enough, and cardiac function, then brain function, have been brought into the picture. Various kinds of coma and vegetative states have proven to be reversible, with physiotherapy and cognitive therapy helping to adapt around the degenerative muscular and neurological damage caused by them.

A recent practice of using low temperatures to slow the metabolic functions of the body have been put to amazing use in 2015 when a teenage boy in Italy drowned and was dead, according to traditional definitions, for 45 minutes, was revived, and through a process that last months, brought back to normal bodily and mental functions (less a leg he lost in the drowning accident). If from a few seconds of lack of heartbeat, death can become so vaguely defined as being beaten back for three quarters of an hour, can we hypothesize prolonging suspended states indefinitely, where metabolic functions are suspended and the body and the mind don't decay?

Cryonics is the study and the practice of this, with companies already offering their services to customers who get into the care of their services after being declared legally dead, but before natural cellular decay can destroy them. They do it without recourse to traditional freezing, which, through the formation of ice crystals, irremediably damages the organs, but to vitrification that substitutes the body's fluids with a solution that congeals like glass at low temperatures.

Applying all the available treatments and progressive rejuvenation as available, a cryonic suspension policy can be seen as a bridge towards more radical options to be developed in the future. Policy is actually an appropriate term, as there is at least one insurance brokerage offering a package that includes in its payout the coverage of a cryonic service for its policyholders.

To be an onion

We were sitting around the dinner table, my wife, three children and I. Starting from another thread of the conversation, my oldest son confirmed that if happened that he ended up in an irreversible coma, a vegetative state, he wanted to be turned off, the machines disconnected. My youngest daughter chimed in that no, she didn't want to be disconnected, but wanted to be kept on even as a vegetable. I could not refrain from asking: "That's OK. But which vegetable do you want to be?" After thinking a bit she said "An onion, or a carrot", and we went around the table confirming in turn what vegetable eventually we wanted to be if in a coma.

The choice of not being revived, of declining extreme and invasive medical procedures that may prolong life a bit but at a very low quality, is a freedom that is spreading now. It is likely to embrace the concept further, generalizing into the right to choosing one's moment to die. The reason for wanting to avoid being powerless and unthinking and only kept alive by machines is altruistic: we want our relatives to be able to move on. If they are there with us indefinitely, they are also as good as dead, locked in a system that is hopeless and useless together with us. (I am not disparaging the efforts that are sometimes fruitful of dedicated relatives that are able to revive those who are not in an ir-reversible coma. And the medical diagnosis is often not clear-cut, which contributes to making this topic especially complicated and emotionally fraught for those who make the decisions after the fact.)

Cryonics, as a consequence, is for an individual to be able to voluntarily contract a service that is relatively simple and cost effective if compared to what a patient in deep coma needs, while unlocking a future for the relatives that is unencumbered by the presence of a person that can't actively participate in it, at least for the time being.

Mind uploading

Obviously the best efforts of science, health and medicine can't still stop the proverbial bus. Accidents will still happen, and will interrupt the trajectory of any life sooner or later. In many of these cases that outcome will be such that the cryonic suspension emergency team won't be able to get to the scene in time, or find remains that are not worth preserving. The solution for successful data recovery in the case of computers is a reliable backup procedure, and there are now teams of researchers working on the ways that the human brain, its neurons, synapses, and any other needed structure that gives rise to the mind could be imaged and preserved.

Functional magnetic resonance imaging is a process that creates a three-dimensional image of the brain, not only recording its geometry but also the firings of the neurons, the activities of the synapses. Its resolution is increasing at an accelerating pace, and is one of the candidate technologies for being able to record and reproduce in a sufficiently detailed manner what happens in the brain to preserve it.

Any backup procedure is only as good as the ability to access and use the data after it gets restored. And restoring a human mind would require another human brain into which to restore it, likely to be impractical given the ethical implications, even if there are already steps towards one of the most hair-raising operations that traditional procedures can design: a full head transplant. An alternative is to actually complete the restore step onto a medium that is different from a biological brain, a support that would not only be able to store but also to execute the brain functions giving rise to the experience of the mind. The eventual successful execution of this would definitely respond to both the Turing test and the feasibility of AGIs, as a restored you in silicon, or whatever else the necessary support would be, certainly would profess to be truly thinking and self-aware, as well as possessed of the capacities of general problem solving of humans.

Kinds of lives

There are many kinds of backups. Those that require the system to be frozen during the process, or others that are continuous while the system is running. And there are also many ways to test the integrity of the data, for example executing a restore and running the system, without a disaster having destroyed the integrity of the original. When you get restored, even if the original is still around and this step is only to test that everything works, it is sure that you'll ask to please be kept around anyways.

Living parallel lives, rather than a series of experiences one after another, is an additional variant that will be made possible by the technologies of mind uploading and restoring in different substrates. Of course the various instances of me (the word is likely to acquire a plural "mes"), will diverge per definition, by having different experiences. These individuals will than have the option to merge the experiences of one another into a single tapestry of multifaceted recollections. A ritual will emerge, even if this merging can be done remotely and in a continuous fashion, of a yearly meeting or once in a decade, to accomplish it in a formal setting, where the identities instantiated would be missing only for two reasons: either the then-privileged solitary choice of traditional unrecorded death, or the journey to the stars.

When the sun goes out

For those who care about humanity, the protestations from others of earthly obligations notwithstanding, space colonization is a must, starting with Mars. Unless we become a multiplanetary species, our future is imperiled by an extinction level event that makes our only home planet uninhabitable. Statistically, for example, a very large meteor strike is to be expected every few tens of millions of years.

Similarly, interstellar travel is also a necessity. Much further out, in a few billion years, the Sun is also going to radically change, expanding into a red giant star that is going to swallow all the space from its current volume out to beyond the orbit of Mars.

The traditional human form is very well adapted to living on the surface of a planet with atmosphere rich in oxygen, abundant liquid water available, and about 1 g of gravity. Well, not surprisingly, just like Earth. It is at the same time exceptionally maladapted to thriving in other environments, such as one with no atmosphere, no water, temperatures of −200 C° or thereabouts, and no gravity, i.e. space. In the middle, gravity yes, water frozen, atmosphere yes but thin and poisonous… is Mars, the only planet we know of as of today colonized by robots: ours.

When possible, and the creativity, dexterity and problem solving ability of biological humans will be available in forms that are better adapted to space, whether we'll still call them robots or they will be practically and legally humans in robotic substrates, the true colonizers of space are not going to be meatbags in tin cans.

There is also going to be a rapid process of miniaturization.

As long as computational and operational capacities are preserved, in terms of propulsion the smaller an object, the lesser its mass, the easier it is to accelerate. Nanoscale thinking robotic humans smaller than a grain of dust are going to be propelled by the billions by laser beams to speeds that will be close to that of light to spread out in onion skin layers across the universe in spheres that initially centered on the Sun will soon start spreading out from other centers too, intersecting, interfering, like ripples in three dimensional waves in the continuum of space.

Imperceptible to any technology less advanced than themselves, these waves are going to proceed and progress, building further instances of trillions and quadrillions of minds to swarm in the galaxy, crossing its perimeter in only a few hundred thousand years, and arriving to Andromeda in little more than two million years. The adventure of knowing the deep cosmos will have started, with billions of other galaxies ahead.

Adaptation, individual and self-perception

At the start of the mind uploading, merging, and restoring process very early on the questions of compatibility will come up. How far can the divergence of experiences go until a full re-merging won't be possible? Or the reverse, how close two separate individuals must be, in love maybe?, until they can attempt the process of merging their experiences and minds? The partial process is going to be explored and used more and more widely, both among different instances of the same individual and among the groups of friends, lovers, diplomats, and co-workers that feel the need of a close understanding and collaboration.

With the digitization of the identity, and the instantiation in further substrates, the question of the origin of one, whether human or AGI, will be quickly moot. Humans in alternative form will immediately take advantage of the new opportunities of introspection and constant upgrading. AGIs will acquire human rights and duties in order to participate and make humans participate in the global civilization as equals.

The very concept of individual will blur, to be more usefully defined by the needs of a given challenge that could require the temporary merging of minds and pooling of resources of dozens, thousands, or billions of people. These organizations, we would have called corporations, governments, and societies, are going to be able to negotiate the complex arrangements of inputs and outputs that such a situation will imply.

The galaxy faring clouds of lightspeed explorers that are our descendants will live very different lives than we do, and it is going to be hard for them to identify themselves with us. Looking at the limitations in time, space and opportunities of what we could do both as isolated individuals, our unavoidable fate spares the feeble attempts at communication we achieved, and as groups in coordinating our actions, with all the indications of conflict being witness to how bad at this we were, these entities will reflect and consider how astonishing it is we are connected, how unlikely, improbable or maybe even impossible. The deniers of evolution amongst them will ignorantly point to our radical differences to claim that indeed, whatever they will decide to call their community of species and civilization, they are not human, but we know they will be.

The Simulation Argument

Formulated by Nick Bostrom in 2002 the Simulation Argument says that one of the following three statements must be true:

1. We are the first and only advanced technological civilization in the universe
2. When the technology is available, advanced civilizations choose not to simulate universes

3. Our universe is a simulation and we are living in it.

The simulation argument can be used as an ontological tool. Do you want to maximize the probability of living in a reality that is not simulated? Try to prove that we are alone in the universe. Do you think that the Keplerian revolution of abandoning the idea of a fundamental reality is a worthy achievement? Try to design an ethical system that allows the responsibility to simulate universes containing advanced intelligence.

An interesting question arises from the combination of the "AI getting out of the box" issue described previously and the simulation argument: if a universe is a simulation and contains AGIs, are these going to be successful in convincing the simulators to bootstrap them onto the lower "realer" plain of reality?

A guarantee: the path will never end

Goedel's theorem is a profound mathematical result. Originally a response to a challenge to show mathematics complete, it achieved the opposite, proving that no formal system can be complete, as it will always contain statements that are undecidable, and consistent, as it will always contain statements that are contradicting each other.

The philosophical and epistemological implications of this are staggering. Science will never be complete, based on mathematics, and our exploration of the universe will never be complete, being able to observe systems that are built on phenomena that can be described only through a formal language that includes new elements beyond those already used, in order to account for them.

Deciding what to do, where to go, and how to think about the world in this sense shapes not only our understanding of it, but gives rise to different languages to describe it, to alternative complementary maps of reality.

Hic sunt leones? The porous map of reality

Extending Goedel's results in the '80s it was proven that even taking into account the undecidability of certain classes of statements, for any formal system there is a class of statements that are true, but for which there is no finite countable number of steps that form a path from proven ones. And that the number of these unprovable truths vastly exceeds the statements in the other classes of statements accessible from the given formal system.

Once again these results reshape our understanding of reality. Not only the choices we make about what language to use give rise to microscopes and telescopes, tools to explore, interpret and understand the world that guide us in different directions. Regardless of the direction, we are only bound to catch a sliver of the reality that is vaster than we can ever encompass at any moment. What are the boundaries of those maps of reality that we can weave with our mathematics and science? How do the tricks of incorporating undecidable statements and extending our formal systems play together with doing the same with a chosen set of unprovable truths? How will science evolve to tackle these realms, huge swaths of reality that we may at one point think are beyond its grasp? What will the world that we design through our continued exploration, this new reality, look like? Speculating about this at this point in time may be fruitless, and it is going to be the task of a human-AGI hybrid civilization to continue the adventure.

ARE YOU ONE OF US?

The purpose of this book is not to make predictions about what will be the year in which artificial intelligences will present themselves to be reckoned with. In fact, wether happens within twenty years, as many experts expect, or in two hundred, the effort itself to predict what this event may involve and the possible steps to prepare ourselves represent a great value for humanity.

So, if you arrived to read up to this point, at the end of the book, my hope is that you have acquired an active attitude, looking to the future with open eyes, counting on our collective ability to realize it.

It is essential to spread the knowledge among non-specialists about the challenges that artificial intelligences place on us. Being one of us means speaking, writing, and debating to lay a solid foundation for the next phase of our global civilization.

SOURCE OF ILLUSTRATIONS

Figure 1: my favorite dinosaur, LadyOfHats, OpenClips
Figure 2: "Wheat and chessboard problem" Wikipedia
Figure 3: "Time series" Wikipedia
Figure 4: author's image
Figure 5: author's image
Figure 6: Ray Kurzweil's presentation at H+ Summit @ Harvard in 2010
Figure 7: image from the movie Forbidden Planet
Figure 8: photo by D-Wave Systems
Figure 9: photo by Steve Jurvetson
Figure 10: "Singularity (mathematics)" Wikipedia
Figure 11: photo by Nigel Ackland
Figure 12: photo by Emotive

ACKNOWLEDGEMENT

I want to thank Massimo Temporelli who has insisted on the inclusion of this book in the Microscopi series by Hoepli, my Italian publisher. Without his support and encouragement I could not have imagined writing it.

Most of the book was written in Cuba, and the patience of my wife Diana, who allowed me to organize my time for a creative but methodical work, was unparalleled.

For decades I had a constant inspiration from the conversations with Gianni degli Antoni and his mentorship was fundamental for the development of my thought.

The people who directly or indirectly gave stimulus, feedback and influenced the structure and contents of the book with their work: Ray Kurzweil, Marvin Minsky, Neil Jacobstein.

ABOUT THE AUTHOR

David Orban

David Orban is an investor, entrepreneur, author, keynote speaker, and thought leader of the global technology landscape. His entrepreneurial accomplishments span several companies founded and grown over more than twenty years.

He is the Founder and Managing Partner of Network Society Ventures, a seed stage global investment firm focused on innovative startups at the intersection of exponential technologies and decentralized networks.

David is the Founder and a Trustee of Network Society Research, a London-based global think tank present in over 40 countries, creating a vision and analytical tools to allow individuals, enterprises and the society at large to deal positively with the unstoppable transformation to a world based on decentralized exponential technologies that are disrupting the traditional centralized and hierarchical functions of governments and corporations.

He is also on the Faculty of and Advisor to the Singularity University, based in Moffett Field, California, an investor in the Singularity University Labs Accelerator Fund (now SU Ventures). SU is an interdisciplinary university whose mission is to assemble, educate and inspire leaders who strive to understand and facilitate the development of exponentially advancing technologies in order to address humanity's grand challenges.

As a sought-after speaker, he has given over 100 keynote addresses and speeches around the world for organizations including Abbvie, Cisco, Oracle, Roche, Ernst & Young, Accenture, Gilead, TEDx Academy, ENEL, Intesa, Banca Sella, Mediolanum, Alphabet, Internet Advertising Bureau, European Foundation for Management Development, GALA, Login and H-Farm.

Born in Budapest, Hungary, he studied Physics at the University of Milan and the University of Padua.

David's work cuts across the limits of specialization to contribute to a new renaissance. He explains, "My vision is at the crossroads of technology and society and their co-evolution." His personal motto is "What is the question I should be asking?" This concept is his vehicle for accelerating cycles of invention and innovation in order to build the brave new world ahead.

He can be contacted via email:
david@davidorban.com
and on his website on davidorban.com